ESSENTIAL GUIDE TO

ECOLOGICAL GARDENING

Quarto.com

First Published in 2026 by Cool Springs Press, an imprint of The Quarto Group,
100 Cummings Center, Suite 265-D, Beverly, MA 01915, USA.
T (978) 282-9590 F (978) 283-2742

EEA Representation, WTS Tax d.o.o.,
Žanova ulica 3, 4000 Kranj, Slovenia.
www.wts-tax.si

30 29 28 27 26 1 2 3 4 5

ISBN: 978-0-7603-9829-6

Digital edition published in 2026
eISBN: 978-0-7603-9830-2

Library of Congress Cataloging-in-Publication Data available.

Design and Page Layout: Samantha J. Bednarek, samanthabednarek.com
Cover Images: Shutterstock (front), JLY Gardens (back top), and Janet Davis (back bottom)
Illustration: Melissa Bryant, melbryarts.com, on pages 42, 54, 69, 123, 124, 125, 147, and 148

Printed in Guangdong, China TT092025

ESSENTIAL GUIDE TO

ECOLOGICAL GARDENING

Techniques and Know-How for Gardening with Nature

CONTENTS

MEET *the* AMERICAN HORTICULTURAL SOCIETY

FOUNDED IN 1922, the nonprofit American Horticultural Society (AHS) is one of the most respected and long-standing member-based national gardening organizations in North America. The society's membership includes more than 22,000 aspiring, new, and experienced gardeners, plant enthusiasts, and horticultural professionals, as well as numerous regional and national partner organizations.

Through its educational programs, awards, and publications, the AHS inspires a culture of gardening and horticultural practices that creates and sustains healthy, beautiful communities and a livable planet. AHS is headquartered at River Farm, a 27-acre (1 ha) site overlooking the Potomac River that is part of George Washington's original farmlands in Alexandria, Virginia. The AHS website is www.ahsgardening.org.

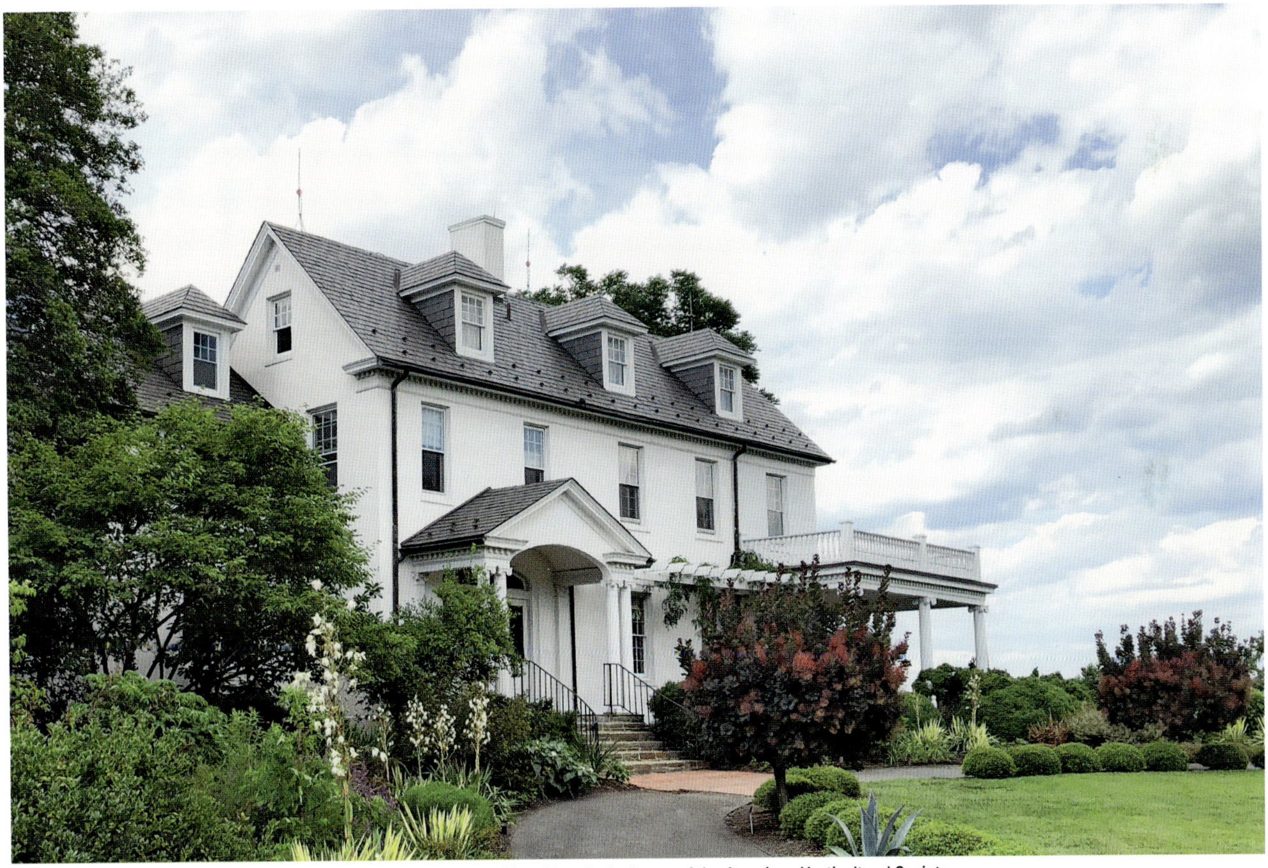

Twenty-seven-acre River Farm is the home of the American Horticultural Society. It's located on part of George Washington's original farmlands overlooking the Potomac River.

INTRODUCTION

IN A NATURAL, SUSTAINABLE, ECOLOGICAL GARDEN, humans are not wholly in charge. Ecological gardening is a collaboration with nature.

You may find that statement difficult to read, as we're used to thinking about our place at the top of the food chain. You may also find the statement surprising at the start of a book about gardening, but it's a fact every ecological gardener needs to learn.

Humans belong to nature, but we're just one element here. The larger natural community comprises, in small part, millions of animal species, hundreds of thousands of plant species, 3,500 species of bees in the United States alone, and at least a trillion species of microorganisms that make life on Earth possible. Ecological gardening puts an emphasis on supporting the whole ecosystem in our designs and decisions.

This is somewhat of a beautiful rebellion from the direction that much of the gardening industry has taken. Ecological gardening is a diversion from the mainstream superhighway of flashy, newer, and faster varieties, techniques, and technologies to a more meandering backroad of observation, care, and thoughtful cultivation. It's a forward-thinking look at conserving and protecting water, improving soil, and reducing our reliance on other resources, which considers not just our needs but those of our neighbors—human and otherwise.

A garden created using an ecological approach is a treasure for you and for all living beings in your community. It's likely that you've come to this realization on your own, and that's why you're holding this book.

Gardens built for resiliency are created in careful partnership with nature.

When you commit to thoughtful garden design and care, everyone benefits—not just people.

There are many reasons why this may be the garden book you need at this time:

- Your desire to bring butterflies and hummingbirds into your garden
- A need for something other than a traditional lawn to help mitigate the effect of significant rainfall or drought on your home and neighborhood
- Your memories of tasting the season's first homegrown tomatoes from your grandmother's garden and your interest in recreating that family tradition in a chemical-free way
- An interest in environmental concerns and wanting to create a refuge for your wild neighbors amid the human disturbance of housing development and habitat reduction
- A way to contribute to your family's and community's resilience

Everyone comes to gardening from a different place. Whether you've never held a trowel before or you're known for your prize-winning petunias, there's room in the ecologically conscious garden for us all.

ABOUT THE *AHS ESSENTIAL GUIDE TO ECOLOGICAL GARDENING*

This book is designed to deepen your understanding of what it means to work with nature to create a beautiful and functional outdoor space. Apply the advice from this book to your existing garden layout, or use it as a guide to create a whole new, beautiful landscape. The emphasis here is on function, from an ecological sense, and can be translated to ornamental and edible gardens.

You'll learn to see the garden as a habitat for all kinds of living creatures, including yourself and your family. In viewing it as a whole ecosystem, you can connect the dots between how your actions in your own space affect your neighborhood and spaces beyond. The importance of relationships, between other people and with the rest of the ecosystem, becomes evident.

In the first chapter, you'll be introduced to the philosophy and essential nature of ecological gardening and how this relates to garden-world buzzwords like "organic," "permaculture," and "sustainability." You'll be empowered to develop your own ethos for what ecological gardening means to you and to determine guiding principles for how you'd like it to look in your space.

Rather than offering a strict how-to directive for constructing an ecological garden, this book unfolds as a guide to combining aesthetics and ecology. As you work through the chapters, you'll better understand the vital biodiversity of garden life. You'll be asked to consider the plants most suited to your region, which might mean replacing the showiest new cultivar in the catalog with a stunning native perennial. You'll be reminded that nature takes its time and that gardening alongside nature requires patience. You'll have the chance to question some of the things you might believe about what plants need to thrive and what gardens need to be beautiful.

A whole chapter is dedicated to our precious water resources. Water movement and capture, its responsible use, and the issues of having too much or too little water are all essential topics in the ecological garden. You can apply this chapter's lessons to work with and improve the conditions in your own yard.

Throughout these pages, and particularly in chapter 6, ecological gardening techniques are explained. Beginning with how to prepare an area for a new garden, into plant propagation, and then how to maintain the garden, these are tasks every gardener will perform, with a focus on how to do so in line with the natural community. The exploration of nature-friendly garden-maintenance practices reveals the unfortunate news that weeding is still a chore, even in the ecological garden, but offers the bright side, which is that "weeds" have a place in the garden too. In examining the role of fertilizers and soil amendments, and the difference between the two, you have the chance to question the inputs you may have used or been instructed to use.

This book's final chapter expands ecological gardening concepts into a more comprehensive picture of an ecological life. It's hard to separate the two when you further your understanding and embodiment of all of life's interconnectedness. You'll dig deep into plastic use, how and why to build a circular economy in the garden and outside of the garden, how to support your neighbors, and whether you need so many material things. The book comes full circle with further evidence of the importance of building relationships.

WHERE YOU FIT IN

Because humans are not actually in charge in the natural world, you may look at ecological gardening as giving up your need to control outcomes. The full premise of this book revolves around working with nature rather than trying to circumvent nature. This is a tall task in our driven, success-oriented society. What you may find in these pages is that it's also an enjoyable task. There is something special that happens when you can let go, and in the case of letting go in your garden, the results can be beautiful, both aesthetically and on an ecological level. There's a slowness to the process that you may have missed when you were growing purely for the sake of the showiest flowers, the earliest tomatoes, or the most abundant herbs.

It's possible that through the practice of ecological gardening you will learn more about yourself and your community. Tuning into nature in this way can even be considered somewhat of a meditation. Your job now is to pay attention to your weather, your climate and how that's changing, your soil health, and the plants and organisms that thrive in the landscape. You're also asked to tune into yourself and consider what you want to grow and why. You are part of the ecosystem too.

This book is your permission to let go of the thing that you "need" to grow or the way your garden "should" look. Partner with nature, and together you can take it from here.

Working with nature instead of trying to circumvent it can yield beautiful results, both aesthetically and ecologically, as is evident in this autumn scene of American beautyberry (*Callicarpa americana*) backed by switchgrass (*Panicum virgatum*).

CHAPTER 1

THE ECOLOGICAL *Gardening* APPROACH

WITHIN OUR NATURAL COMMUNITY, humans have an outsized role in shaping systems and shifting directions. A holistic view of our place in the natural order is not often prioritized over political and financial gain. Ecological gardening calls us to set those aside in favor of the bigger picture of our human relationships, our plant and animal (and microbial) neighbors, and the order of things downstream from our yards. Our decisions, cumulatively, shape all of their lives as well as our own.

As ecological gardening looks above and beyond simply what your yard can gain in this transaction, details string together to make interconnection obvious. After a little time spent in observation in the garden, you'll notice the ladybugs (beneficial insects) eating the aphids (pest insects) and will realize that, had you sprayed the aphids with a pesticide, the ladybugs wouldn't have arrived. As the ladybug population grows, you'll find the birds having their fill of those insects. And now your bird population is growing, and you can witness their nest building and song singing right outside your door. The bigger picture is that ecological gardening provides space for all of these natural neighbors to thrive.

Stepping away from the need to control every element of the garden gives you as the gardener more space to thrive as well. Think of the money, time, and physical effort that can be spent in a garden fertilizing, tilling, watering, and otherwise maintaining unreasonable and unnatural aesthetics. There's so much that you cannot control, from precipitation to temperature and whatever your neighbors are spraying in their yard. Supporting your garden's natural processes inherently makes it, and your community, more resilient in the face of these variables. A bonus to ecological gardening is that, while it still requires effort, this approach can be easier on your back and your bank.

WHY *Ecological* GARDENING

There are a plethora of reasons to garden with an eye toward diversity and resilience, not the least of which is beauty.

REGARDLESS OF YOUR personal motivation behind ecological gardening, the impact is great for humans, the natural environment, and the built environment. Consider the benefits:

Stormwater management. The need to manage stormwater is more important now than ever for several reasons. More of the United States is seeing significant rain events, sometimes lasting several days and sometimes covering a whole season. These are being called *hundred-year floods,* but in some cases, they're happening just a few years apart. More of our land mass is being paved over, leaving less area to soak up the water falling on and moving across the ground. A healthy garden, with healthy soil, acts as a sponge to absorb water and slow its movement across the landscape. One study found 4 inches (10 cm) of rain per hour can infiltrate a lawn, but a landscape with trees can soak up 10 inches (25 cm) per hour.

The plant leaves themselves, especially in a dense tree canopy, delay the water from hitting the ground and even give it a chance to evaporate. Large evergreen tree canopies, in leaf year-round, can capture thousands of gallons of rainwater this way.

Ecologically managed gardens' stormwater diversion reduces the burden on municipal sewer systems and on fragile landscapes and waterways while also filtering pollutants from the runoff. Water soaking into the ground replenishes aquifers, the groundwater reserves that millions rely on for home use.

Temperature control. You've probably heard of the urban heat island effect: Cities heat up more rapidly and more intensely than the surrounding countryside because the buildings and streets act as heat sinks. With all the benefits that vegetation and ground cover offer, healthy gardens counter the heat island effect. You'll learn more about the role of plants in the heat island effect in chapter 4.

While that's a big-picture benefit, more specifically, trees offer temperature-regulating benefits to your home and your neighbors'. Sited to the west, east, and southwest sides, deciduous trees shade buildings in the summer while allowing sun to reach them after their leaves fall. Evergreen trees to the northwest block some of the harsh winter wind. As your garden matures, the trees' benefits increase. A typical residence with an evergreen windbreak can save as much as 50 percent on a winter heating bill. Trees or shrubs shading your air conditioning unit allows the unit to operate more efficiently, saving energy, wear and tear, and operational cost too.

Pollution control. All plants, and especially trees, are essentially solar-powered air filters, adept at removing pollutants from air and water.

Through the process of photosynthesis, plants use sunlight's energy to break down water and carbon dioxide. They use the resulting sugars as energy, and they release oxygen back into the atmosphere.

Plants remove carbon from the atmosphere, and the roots deposit carbon into the soil. Scientists estimate that forests hold 45 percent of all terrestrial carbon. Your home garden is smaller than a forest, sure, but your home garden coupled with every home garden of every reader of this book will cover some ground. Perennial and native plants, which will live for years without requiring chemical inputs or major soil disturbance, are especially suited for carbon sequestration.

As water moves across surfaces, it picks up natural and synthetic particles along the way and carries them along. This includes everything from eroded soil and excess fertilizers to motor oil and microplastics. Some of these particles are filtered by the soil so they don't reach waterways.

Plant roots also take charge in filtering pollutants from water. Called *phytoremediation*, the roots soak up the water and some of the pollutants it carries, removing toxins from the soil.

Biodiversity. You already understand that an ecological approach to gardening considers all life, not just human life. This garden provides a corridor for wildlife, insects, and migrating birds and butterflies. Depending on your location and the plant species you include, your garden could even become an essential breeding and feeding ground for monarch butterflies. At the time of this book's writing, monarchs are on the verge of being listed as a threatened species by the United States Fish and Wildlife Service, meaning we need all hands on deck to support their continued existence. You'll read more about wildlife, bird, and insect forage and habitat in chapter 3.

A pleasant space. Do not underestimate the emotional benefits of a healthy garden. The act of being outdoors in the garden is beneficial. Many studies have proven high emotional wellbeing among gardeners, not to mention the physical well-being that comes from your garden tasks. Even just having exposure to green space has been linked to better health. Is it better to look out your window at a neighbor's driveway or at a flower-studded hedgerow visited by birds and squirrels? The more pleasant your neighborhood becomes, the more you and your neighbors will want to spend time outdoors in it.

Higher property values. It's hard to quantify ecosystem services, but money talks. Real estate professionals know that properties with trees and landscaping have a higher value.

Less noise. While urban noise pollution is hard to escape, vegetation in the form of windbreaks or hedgerows can reduce audible noise by nearly 50 percent. (You'll read more about hedgerows in the next chapter.) Additionally, you won't need your leaf blower to manage an ecological garden, and you'll have less lawn to mow, so this management style inherently generates less noise.

Trees offer many benefits to the ecological landscape, including shading your home to reduce cooling costs, controlling erosion, and providing food and habitat for endless creatures.

Roadside plantings can reduce noise, block unsightly traffic, beautify your neighborhood, and provide resources for pollinators and other wildlife.

Reclaiming a natural history. Before industrialization and colonization ever got hold of the land we inhabit, nature had its own design. Consider the rivers that have been dammed, the streams that have been redirected underground, the wetlands that have been filled in, and the forests that have been clear-cut to make room for our highways, houses, high-rise buildings, and farmland. An ecological garden is an opportunity to reclaim and recapture some of your region's natural history that has been lost or damaged. Your garden's design can bring back remnants of historical features to your property and community.

Personal ethos. For anyone wishing to lessen their negative impact on the natural world around them, creating a garden that helps to support the ecosystem is a worthy pursuit. Ecological gardening is in alignment with many people's desires to simply do better for themselves and those around them.

A BALANCED ECOSYSTEM

The often linear built environment of a neighborhood or subdivision can become more ecologically balanced by including a diversity of native plants in both front and backyards.

THE SUN AND THE RAIN. A garden pest here, a beneficial insect there. Nutrients taken from the soil to grow a tree, nutrients returned to the soil with fallen leaves.

All around you are systems and communities functioning in balance. The ecosystem holds the predator-prey relationship, the nutrient cycle, the water cycle, and other examples of the natural balance of the world. These are closed-loop systems, meaning there's no need for inputs to keep them going and no waste left over.

In a built environment, systems are more often linear than they are cyclical. Think about a restaurant business. It requires inputs of packaged food items, payments from customers, and electricity and water utilities. It outputs food for people, waste in the form of packaging and leftover food, and payments to employees. At the end of the day, we go back to the beginning of the line and start over, consuming inputs that are external to the outputs.

DEPLETION AND ABUNDANCE

When we step in to "improve" our surroundings, we usually do so for the sake of human desires, which may not be in line with the needs of the larger ecosystem. Each environmental change we make has a multifold impact on the ecosystem. We can't always see the effects of our decisions unfold, because the impact doesn't always happen in our neighborhood; rather, many positive and, especially, negative influences become evident somewhere downstream or downwind.

All living things make up the world's biodiversity, from several billion humans all the way down to the microbes that far outnumber us. As we alter our environment, we alter habitat for the living things around us. This can disrupt nature's most basic cycle of animals and plants that rely on one another as hosts and food sources. One small change creates a domino effect.

There are abundant examples of human decisions altering natural cycles.

Consider bats, our ally in controlling insect populations. When increasingly severe weather, housing and industrial development, pesticide use, and light pollution drive bats from an ecosystem, both farmers and everyday people notice their absence.

Bats get a bad rap, but we'd be in big trouble without them and their voracious appetites. Each night, bats can eat up to half of their body weight in insects, and it's possible for pregnant or nursing bats to consume 100 percent of their body weight in insects. For farmers, bats help control insect-pest populations around crops so there isn't a need to rely so heavily on harmful pesticides; for the rest of us, bats make it more pleasant to spend time outdoors without being harassed by mosquitoes. Bats also serve as pollinators, and fruit-eating bats distribute seeds. By tipping the natural balance of the bat population, we all feel the effects.

Though bats often get a bad rap, they are essential predators, consuming many insects that are both human and agricultural pests. Erecting bat boxes and shelters is a great way to provide them with a safe haven.

Everything that happens on land eventually impacts the watershed. Clear-cutting forests impacts the way water flows on the land, increasing both erosion and pollution runoff, causing negative effects for miles downstream.

On the other side of the balance, as a higher concentration of a member of the natural community moves in, they bump out others. This happens in bacterial, insect, land, and aquatic populations, as well as in the plant community.

To illustrate this idea, look at native versus invasive plants. Native plants are those that were growing in a place before European colonization and have adapted to the area's growing conditions for thousands of years. Invasive plants are those that have been introduced in the past several hundred years and cause ecosystem harm.

From Alaska to Alabama, nonnative plants introduced by humans are taking over landscapes. Most of these plants have been introduced intentionally, such as English ivy, which was brought to the United States by European colonizers as an evergreen ground cover in the early 1700s; some plants arrive unintentionally, including Canada thistle, which is thought to have been introduced from Europe in the 1600s in seed shipments. These plants are bullies, outcompeting the plants that traditionally thrived where they now grow, disrupting forage and habitat for native wildlife, and sometimes wreaking other havoc.

Not all nonnative plants become invasive, of course. Food gardeners grow tomatoes, which are native to South America, without these plants taking over. You'll read more about native and invasive plants and their role in your garden in chapter 4.

DOWNSTREAM CONSEQUENCES

One more area of ecological balance worth noting is the connection between land use and water health. Each of us lives in a watershed: an area from which precipitation and runoff flows to creeks, rivers, and larger bodies of water. If you live in the Chesapeake Bay watershed, the runoff resulting from your neighbor's efforts to wash their car in their driveway flows into the storm sewer and eventually makes its way all the way to the Chesapeake Bay. Everything that happens in your watershed impacts the water quality and availability of the end body of water. This is related to ecological balance because everything that happens on land eventually impacts the water. When a forest is cut, plant and animal habitats are changed and so is the way waterflow interacts with the land. The trees held the soil to prevent erosion, the roots helped filter pollutants, the rich soil organic matter allowed more water to infiltrate the ground rather than run straight across it. With the trees gone, sediment, more pollutants, and heavier waterflow reach tributaries faster, which disrupts aquatic life and can lead to flooding of communities in the path.

These are just a few examples of the delicate balance within our natural cycles and the impacts that result from disrupting the equilibrium. As an ecological gardener, you play a key role in maintaining the balance in your garden ecosystem. It's up to you to contextualize how your gardening decisions affect other living beings down the line.

Ecological Gardening Companions

The values behind ecological gardening are similar to those of other gardening methods. You can incorporate parts of these into your personal ecological gardening principles too.

ORGANIC

Before 2002, "organic" food meant food produced without chemical sprays and fertilizers, but since the establishment of the United States Department of Agriculture's (USDA) National Organic Program (NOP), the word has taken on a new meaning. The USDA governs the use of the word *organic* in a commercial sense. For something to be labeled organic, it must meet the standards of the USDA NOP and be properly inspected and certified each year. This is a complicated process and not one that a backyard ecological gardener is likely to undergo.

Just because your garden isn't a USDA Certified Organic space doesn't mean you can't borrow from USDA NOP guidelines. In fact, these guidelines hold many basic principles of ecological gardening. Read all of the guidelines at www.ams.usda.gov/rules-regulations/organic.

A common criticism of certified-organic agriculture is that it doesn't go far enough in holistic ecosystem protections, such as permitting large-scale tillage and lacking in human-resources considerations for farmers. As an "uncertified organic" ecological gardener, you can fill in the gaps with your own, more stringent principles.

PERMACULTURE

Based on Indigenous and traditional ecological knowledge, the practice of permaculture was promoted in Australia beginning in the 1970s. Permaculture design mimics the diversity, functionality, and resilience of natural systems, starting with the home and human ecosystem and working outward into the community.

Though permaculture has had a rocky history, thanks to some practitioners advocating for the intentional planting of invasive plants, many current practitioners have embraced the use of native plants and a more thoughtful sharing of Indigenous knowledge. Every part of a permaculture design is intentionally chosen to provide a service or fill a role related to natural systems.

Permaculture has twelve principles, which apply to both garden and community design:

1 **Observe and interact.** Take the time to learn about your property and community before making changes to your garden and lifestyle.

2 **Catch and store energy.** Use what nature provides.

3 **Obtain a yield.** Essentially, employ plants and systems that are productive.

4 **Apply self-regulation and accept feedback.** Let your system be self-regulating when possible, and make adjustments based on your observations.

5 **Use and value renewable resources and services.** Limit consumption, and reduce your fossil fuel reliance.

6 **Produce no waste.** Get as close to a closed-loop system as you can.

7 **Design from patterns to details.** Think big picture, and then focus in closer as you go.

8 **Integrate rather than segregate.** Work with nature, rather than trying to outsmart nature.

9 **Use small and slow solutions.** Create local, closed-loop systems and smaller, easier-to-manage gardens.

10 **Use and value diversity.** Biological and cultural diversity are key to resilience.

11 Use edges and value the marginal. In reading about the diversity of life-forms in your garden, you'll learn that the places where two ecosystems meet is the most biologically diverse.

12 Creatively use and respond to change. Accept that nature will shape your garden space, and be willing to work with and adapt to these changes.

There is no shortage of permaculture information available to you, if you wish to incorporate these principles into your ecological garden design.

SUSTAINABLE

What a sustainable garden looks like in practice varies widely. This is because, while there are a lot of people, organizations, and brands using the word *sustainable*, there's no one agreed-upon definition of sustainability.

The American Horticultural Society looks at sustainable gardening as combining organic gardening practices with resource conservation. This is a component of ecological gardening, for sure. Looking at the bigger picture of "sustainable," the word refers to a management method that doesn't deplete or damage resources. Of course, this is a component of ecological gardening too.

Clearly, they're related, yet ecological gardening goes even further than sustainable gardening. Rather than simply not degrading the ecosystem, ecological gardening supports it.

Looking at sustainability from a systems standpoint, the concept is often referred to as being a three-legged stool: people, planet, and profit. To apply this to your garden, you would aim for a space that's:

- Pleasant for people to use and reasonable for people to maintain
- Not harmful to the environment
- Functional, in the sense of ecosystem services that it provides

Even in a single raised bed, you can practice permaculture techniques by following its twelve core principles.

REGENERATIVE

Like *sustainable*, the word *regenerative* has different definitions, depending on who you ask. In general, regenerative landscaping or farming refers to a practice that improves, rather than simply maintains, the overall health of the land.

Now and then, an organization will formalize their own definition, as the California State Board of Food and Agriculture did in 2025. The full-page definition begins:

"'Regenerative agriculture,' as defined for use by State of California policies and programs, is an integrated approach to farming and ranching rooted in principles of soil health, biodiversity, and ecosystem resiliency leading to improved targeted outcomes. Regenerative agriculture is not an endpoint, but a continuous implementation of practices that over time minimize inputs and environmental impacts and further enhances the ecosystem while maintaining or improving productivity, economic contributions, and community benefits."

At its core, "regenerative" is certainly ecological, and finding regenerative growing examples may be useful as you build your own ecological gardening practice.

TAKE *Your* STAND

Regularly reflect on your reasons for gardening with a focus on ecology. When challenges arise, doing so will keep you pointed in the right direction.

KEEPING A 100 PERCENT ecological garden is a challenge, to say the least. While you can make many gardening decisions—even most gardening decisions—based on what's best for the world around you, there will be times that you have to accept a less-than-ideal choice. Plastics are ubiquitous, gardening in close proximity to others means their landscape management invariably affects yours, and living a busy life in modern America, sometimes you need to just accept an easy solution at hand.

Here, take a moment to reflect on your reasons for pursuing the ecological garden path. Related to any of the benefits of ecological gardening (on page 228), pursuant to your personal green-living interests, or for some reason this book doesn't even consider, let's line up your principles.

LIVING YOUR VALUES

Start by defining what ecological gardening means to you. Let it be your personal garden mission statement. As you face decisions about your garden, measure your options against this statement. Perhaps one decision is more closely in line with your concept of an ecological garden.

ECOLOGICAL PRIORITIES

What does ecological gardening mean to me?

PRIORITY	PRINCIPLE	IN ACTION
	Conserving water	
	Supporting pollinators	
	Providing wildlife habitat	
	Promoting plant biodiversity	
	Beautifying the yard	
	Remediating and building soil	
	Mitigating climate impacts	
	Reducing waste	
	Reducing fossil fuel inputs	
	Growing food	
	A hands-off approach	

Don't forget to give yourself some grace in designing your ecological garden. Make decisions based on your personal needs as well as the needs of the community around you (both human and nonhuman).

With this in mind, set some priorities. What ecological concepts are most important to you? How do you want to commit to putting these into practice?

Use the table on the facing page to outline your ecological gardening principles. You'll be reminded of them as you read through this book.

In setting these priorities, know that they all work together on some level. The progress you make in one value set will invariably have benefit on other values. Promoting plant biodiversity, for example, supports pollinators, provides wildlife habitat, and beautifies your yard.

Likewise, some of these principles will compete with others. If growing food is a top priority, you may choose to install an irrigation system using plastic drip irrigation components. (Plastic is the only option.) In this case, prioritizing growing food would take away from your ability to reduce fossil fuel inputs. There's give and take in the decisions you'll be making, which points to the importance of establishing your ecological gardening principles before digging in.

BUT I "SHOULD" . . .

Before the pressure of what you or others think you "should" do comes bearing down, give yourself some grace. Look at the world around you and realize that any act of ecological gardening is against the grain. There is no perfect system here, and you're not going to hit every ecological gardening benchmark. Do your best, using your principles as guideposts.

Going Peat-Free

Opting not to use peat moss in your gardening practice is an important step in ecological gardening. The peat used in many potting mixes was formed by layers and layers of sphagnum moss growing in bogs. As the moss grows, the lower layers partially break down, but can't completely decompose thanks to the highly acidic water in the bog. The partially decomposed moss builds up over time, eventually producing a thick layer. Though peat has long been a horticultural staple, there are two primary reasons to avoid it.

First, peat harvesting is not sustainable and very destructive. Most bogs only produce 1 millimeter of peat each year, meaning the thick peat layers we see today took thousands of years to develop. If we keep harvesting peat at the rate we do today, we'll eventually run out, destroying valuable habitat for the creatures and plants that evolved there.

Secondly, peat sequesters carbon. Peat bogs today store twice as much carbon as all the forests in the world put together, and they remove 0.37 gigatons of carbon from the air each year. When we harvest peat for potting soil, we stop bogs from sequestering carbon and release the carbon they have already stored.

Peat bogs are special habitats that host a diversity of unique plants, including this carnivorous pitcher plant in bloom.

A COST-BENEFIT ANALYSIS

IN ECOLOGICAL SYSTEMS, cost refers to more than the financial price tag. Money matters, of course, and it is a primary constraint in how you design and install your garden. Alongside that is the environmental cost, or benefit, of your decisions.

FINANCIAL COST AND BENEFITS

Starting with the financial aspect, an ecological garden may be more expensive to install than a garden of the same size that's not as focused on the environmental costs.

You may find yourself sourcing plants from a local native-plant nursery to start out, which would cost more than picking up mass-produced bedding plants from a big-box store. If you're making changes to the topography of your landscape, such as installing berms and swales, you may have the cost of renting equipment or paying for labor. You could choose to work with a garden designer, and finding someone knowledgeable in ecological design may be more expensive than a more conventional garden designer.

Ecological gardens may have greater installation costs depending on the level of site preparation needed and plant sourcing, but over time, they require fewer inputs than traditional gardens, leading to reduced long-term investments in both time and money.

Ecological gardens benefit so many creatures beyond the humans cultivating them.

While these startup costs are greater, over the life of your garden, you might end up spending less. Assuming you're installing native plants and perennials specifically suited to your climate (see chapter 4), these will require fewer inputs to maintain them. They'll be long-lived, so you won't spend more money on replacements in a few years. You're building relationships with other gardeners to trade seeds and propagated cuttings, saving money you'd otherwise spend at a nursery. The healthy soil you're building will reduce or eliminate fertilizer costs. The beneficial insects that call your garden home will stay on top of the pest insect population, saving the need for pesticides (even naturally derived, organic pesticides). If you're replacing your lawn with an ecological garden, you're saving money on gas and maintenance for your lawnmower or the cost of hiring someone to do this work for you.

TIME AND ENERGY COST AND BENEFITS

While money comes and money goes, time and energy are costs for which we have no replacement. Starting any garden is an enjoyable activity, but a time- and energy-intensive one. If you're moving earth to install the features of your ecological garden, this may require more time and energy than if you're just tilling a patch of ground and putting in readily available plants. Still, getting an ecological garden started doesn't require that much more time than establishing a conventional garden.

In the long run, an ecological garden may save you time. As pointed out above, you won't need to invest time applying fertilizers and pesticides, and you won't need to continually replace plants. You maybe be able to use this extra time and energy to just enjoy spending time in your outdoors space.

Ecological Gardening on a Budget

Gardening should be a hobby or a lifestyle that's accessible to everyone, no matter how much money you're able to invest in the project. In so many ways, ecological gardening is economical gardening. Reducing waste, reusing what's available, and building reciprocal relationships are baked into the whole premise of the ecological approach. With this in mind, there are some ways you can approach a garden with frugality.

Reusing building materials, as this gardener has done in constructing a recycled brick walkway, is a great way to explore your creativity as you keep items out of landfills.

- **Start small.** This advice applies no matter the gardening or landscaping you're undertaking. It's hard to know how much time (and investment) your garden will require, making it logical to start small. Managing a small space, even just a few carefully chosen plants or a well-thought-out corner of a yard, still provides benefits to the ecosystem and to you, as a gardener. Buy or acquire the seeds, plants, and materials you can afford now, and build on this each season. Doing something is better than doing nothing.

- **Revisit your priorities.** Why is it, again, that you're exploring ecological gardening? Come back to your top reasons for taking this path and focus on how to stick to them. For example, if growing food ranks lower on your list than mitigating climate impacts, start with a few native plants rather than buying a whole bunch of different vegetable seeds.

- **Start with seeds.** It's easier to build out a garden using seedlings and saplings, but it's more economical to plant and grow out your own seeds. If you have the know-how, the patience, and the space to do so, give seeds a try. Invariably, you'll grow more than you need, and you can trade the resulting plants with your gardener friends to boost your plant diversity without spending more.

- **Reuse, reuse, reuse.** You'll return to this idea throughout the book. You may be surprised by what you'll find secondhand. Gardening tools, plant pots, hardscaping elements (like deck boards and sidewalk pavers), and more are regularly listed on community boards online, sent to thrift stores, and left at the curb. Using your creativity, you can repurpose many items for garden use.

- **Rely on your community.** Swapping plants, buying bulk compost, and sourcing used materials are three ways your neighbors and fellow gardeners can make gardening more affordable. Look to seed libraries and tool-lending libraries to borrow seeds and equipment to get started. If you need help with garden labor, offer to trade workdays with another gardener so you each have another set of hands (and can learn from one another while you're at it).

Ecological gardens don't have to consist of big, sweeping landscapes. They can be small patches of habitat in gardens of all sizes.

ENVIRONMENTAL COST AND BENEFITS

The greatest reason to embark on an ecological garden installation is the benefits to your life and the world around you. Highlighted throughout this chapter, and continued throughout this book, are the many benefits of ecological gardening. For every action taken in nature, a balance takes place. It's true that not every action you take in your garden will benefit the world around us. You might end up purchasing a bulk amount of compost in a plastic bag or using an excess of water trying to keep new plants alive during a drought. Those actions come at an environmental cost. Look at the overall picture of your ecological garden, though, and it's likely that in the end, it works out to an environmental benefit.

This chapter has offered an understanding of the philosophy and benefits of ecological gardening, the difference between ecological and other types of garden systems, and the big picture behind natural cycles and communities. Placing the rest of the book in this context will further shape your understanding of ecological gardening. In the next chapter, you'll consider your garden's place in your community and explore how to develop the relationships that will foster and support your ecological gardening practice.

CHAPTER 2

BEING *a Good* NEIGHBOR

IN YOUR ECOLOGICAL GARDENING PRACTICE, it's important to be a good neighbor to the natural communities around you. In this chapter, the focus is on being a good neighbor to your *human* community. This looks like building relationships, having an intentional garden design, gardening to match your climate and neighborhood conditions, and learning the art of compromise.

In today's disconnected society, it is possible to avoid human connection, which runs counter to ecological thinking. Nature is the original community, and being a human is not a solo act. Whether your community today looks like a rural township, population 500, or a suburban development with 500 homes, you have the opportunity to build relationships and thrive within it. The garden you're planning isn't just for you, and it won't touch your life alone; it will have a ripple effect across the community.

Your first challenge as an ecological gardener is to know your neighbors. For some of us, this is the largest challenge, but building relationships will pay off in tangible and intangible ways.

Start close to home. Who are your direct neighbors, the ones who share a property line with you? It's okay if you've never talked to them, even if you've lived next door to one another for years. Today's the day. This potentially awkward encounter is the first step in being a good neighbor. From here, keep the conversation going as good neighbors do, even if it's just about the weather.

It's possible that positive encounters with your closest neighbors will open you up to conversations with others. It's not necessary to become everyone's friend, but as you embark on an ecological gardening pursuit, the more people who've had a positive interaction with you, the better. This is especially true in more populated places, where nearby neighbors may have a personal stake in your plans.

While you're getting to know those in close proximity, also think about who else in your community you can network with in relation to gardening pursuits. You're about to spend some time and money at the hardware store and garden store. You can purchase anything you need online, but from an ecological perspective, buying local is better. Depending on your area, "local" could mean a big-box store, or it could mean a multigenerational family operation. Either way, there are actual people who manage and work at these places, and they're probably knowledgeable, if not about ecological gardening specifically, then about the products and tools you're looking for.

Building relationships asks us to slow down and take the time to see other people as actual community members. As reciprocal relationships develop over time, we begin to understand what it means to be a good neighbor.

YOUR NEIGHBORHOOD *Needs This Garden*

Small front yard gardens like this tidy "pollinator patch" are great ways to introduce ecological gardening concepts to neighbors. They are a safe place to start.

WITH THOUGHTFUL PLANNING, an ecological garden is an installation that benefits everyone and everything around you. Refer to the first chapter for the list of benefits that an ecological garden provides (pages 14 and 15).

It's easy to expound upon all the ways ecological gardens benefit neighborhoods. It's less easy to spread the word about those benefits and to entice neighbors to take notice. Ideally, you'll have a cordial relationship with your neighbors before you try to introduce the idea of having an unconventional garden. Wherever you're starting from, though, it's important to just get started. You don't have to go this alone.

REACH OUT TO OTHERS

You're likely not the first person in your neighborhood to wish for a less-conventional garden. Going back to the importance of relationships, have a conversation with neighbors who've already done something like this. Do they have good relationships with their neighbors? How have they navigated this topic in the community?

Your local Master Gardeners group, Cooperative Extension agent (even urban counties have Extension agents), gardening club, environmental council, and other like-minded groups may be helpful in your effort to make connections and educate those around you. They might have a gardening- or sustainability-related event coming up that you could invite your neighbors to, or they might organize one with your urging. Hearing from experts about ecological gardens and their benefits, and being given a chance to have their concerns heard and to ask questions, will put a lot of people at ease.

Homeowners associations, neighborhood associations, and city or county zoning boards all require stakeholder input. Join the board or attend meetings to become engaged with their processes and have the opportunity to offer new perspectives. These are also good groups to connect with your new friends at the gardening and environmental organizations to encourage an exchange of ideas.

Keep your human community in mind when designing and planting your ecological garden. Stay connected to your neighbors and interact positively.

Your Cooperative Extension Lifeline

In this book, the Cooperative Extension is offered as a resource for your garden. Acting as land-grant universities' outreach arm, the Cooperative Extension, or just Extension, is installed in every county in the United States. Extension agents work with farmers, gardeners, homeowners, and just about everyone else to share knowledge and resources directly from the scientists and researchers at their institutions. Extension publishes extensively reviewed, science-based bulletins about every subject from residential water heaters to safe canning guides and the genomic selection of beef herds. For readers of this book, Extension offers localized knowledge, in-depth publications, the Master Gardeners and Master Naturalists programs, workshops, the 4-H program, soil tests, networking opportunities, and more. Best of all, many of Extension's programs are free or low cost. While most Extension agents are on board with the concepts of ecological gardening presented in this book, you may find a few who are not. Their varied backgrounds may introduce differing views. It's your job as the "head gardener" of your own property to be mindful of this and make the best decisions for you and your garden's ecosystem.

Since front gardens are viewed by all, be sure to follow community ordinances. If you hope to change community or HOA rules so they are more in favor of ecological gardens, attend meetings and stay involved.

GO ONLINE

Apps like Nextdoor and neighborhood Facebook groups are now a primary way people communicate within communities. You may be thinking that you came to this book to garden, not to spend more time online. In building relationships, you have to meet people where they are, and very often, they are online.

You are not going to these groups to push an agenda or start an argument. You're there to learn what's important to people around you and craft your message to match that. Refer back to all the reasons why your neighborhood needs ecological gardens (pages 14 and 15). That list contains social, financial, and environmental concerns, and one or more of those will resonate with your neighbors.

PLAY BY THE RULES

Whether governed by a homeowners association (HOA) or the city's or county's planning ordinances, your garden design is likely someone else's business, legally speaking. This bit of advice is more relevant if your home is in the city or the suburbs, as rural gardeners have fewer human neighbors, and fewer of their rules, to contend with.

Regardless of where you live, learn your local laws before you start your garden planning process. While you may not agree with all of them, you either have to follow them or change them. Not doing so is a quick way to put your garden on display in a negative way.

Examples of landscaping ordinances, both in line with ecological thinking and not, include:

- A percentage of the yard that must be permeable, whether using landscaping or porous pavement surfaces
- A list of preapproved plant types
- A minimum height of trees and shrubs at the time of planting
- Limits on fencing types and height, outdoor furniture, and outdoor art

Landscaping rules exist to maintain property values, with an interest in keeping whole neighborhoods within a certain aesthetic, and to reduce public health concerns, such as from rodents, flooding, and wildfire risk. Despite studies showing the benefits of more naturally managed spaces, many communities are slow to change.

Get an up-to-date copy of the landscaping ordinances that apply to your property, and ask questions of your homeowners association, county clerk, or other governing official so you're clear on the dos and don'ts of your landscape. It's helpful to understand the reasoning behind the rules so you can learn how to work within them. You may find that your garden dreams fall well within the guidelines or just slightly outside of them, requiring easy adjustments. In some situations that aren't entirely in alignment with the rules, you may learn there's a simple workaround, such as needing to fill out a request for a landscape plan that your neighbors then approve (which is where your good relationships become essential).

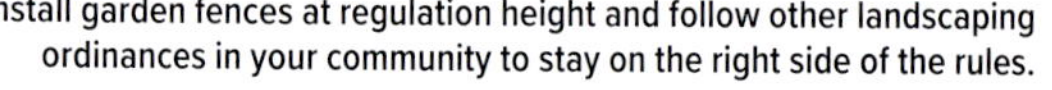

Install garden fences at regulation height and follow other landscaping ordinances in your community to stay on the right side of the rules. 

Certifying your garden as wildlife- or pollinator-friendly habitat and adding a sign to share that certification shows the community your garden is intentional and invites them to ask questions.

Some states, including Florida and Maryland, have laws that prevent homeowner associations and local ordinances from overstepping landscape-regulations boundaries. For example, section 373.185 of the Florida Statutes is known as the Florida-Friendly Landscaping Ordinance, which protects "quality landscapes that conserve water, protect the environment, are adaptable to local conditions, and are drought tolerant." Local entities aren't permitted to tell homeowners they can't have ecological gardens, though they can still restrict landscaping elements. In the summer of 2024, Illinois also passed a law, known as the Homeowner's Native Landscaping Act, which prohibits HOAs from preventing people from planting more Illinois native plants as long as the area is maintained. Change is happening.

In addition to what's on the books, your property may feature easements, such as those allowing access by utility companies, have setbacks for drainage, or involve underground utility lines. Take these into consideration before designing an outdoor space of any kind.

BE NEIGHBORLY

Your work in educating and building relationships with others will never end. As gardens are places of abundance, there are plentiful ways to share the joy of your work.

Share what you can. If your garden includes fruits, vegetables, and herbs, bring some of your harvest to your neighbors, or leave them on a table at the end of your driveway for others to take. Assuming you have plentiful blooms, cut some flowers to share. Small, thoughtful acts go a long way to building bridges in the community.

Certify your garden space. Local and national organizations promoting urban forests, native plants, "green" landscapes, pollinator habitats, and more offer home-garden certifications. These celebrations of your efforts not only make you feel good as a gardener but show the people around you that you have credibility and are taking this work seriously. Choose your certification with thought. If you post a sign touting wildlife or another word that might alarm neighbors, this could do more harm than good. Also, be sure your sign itself meets any neighborhood regulations.

Educate. Gardens provide surprising opportunities to educate others. Along with a sign touting your new certification, provide a QR code for neighbors to learn more about the purpose behind the certification. Post a map of your garden that identifies some of the native plants you've chosen. Invite curious neighbors, including kids, into the garden so you can talk about all that's happening there.

AN *Ecological* DESIGN

A GOOD GARDEN DESIGN is itself a tool for good neighbor relations. As you progress through this book, you'll read about creating habitat for insects, selecting plants for their suitability to your climate, and how to work with water, both too much and a lack of it. The truth is that nature is messy, and working with it can be the same. For everyone living in an urban or suburban area, letting your garden run wild, in an ecological sense, may not be possible. In contending with homeowner associations, planning and zoning guidelines, and neighbors' sensibilities, an actual method to your ecological madness is necessary.

In the practice of landscape design, you're asked to follow rules for the size, shapes, textures, and colors of your plants and structures and the line that these create for the viewer's eye. These design elements make a garden look more intentional than haphazard. It's possible to incorporate them in an ecological garden, as long as the emphasis remains on ecosystem function and harmony. In any garden, spacing is somewhat important, but use nature as your guide: As plants mature, nature doesn't leave a lot of uncovered space between them. Plants tend to be grouped more densely than they would be in a conventional ornamental garden and certainly more than in a typical vegetable garden.

Whether your growing space is in sun or shade, getting to know it before planting a single plant is essential for success.

GET TO KNOW YOUR SPACE

All properties have positive and negative attributes. It's your challenge as an ecological gardener to work with these and to make adjustments to further your purpose. Water movement across the yard, shade and sun, wind tunnels, and existing vegetation and structures can all be viewed as attributes, through the right lens.

Before you can truly appreciate the space you have to work with, you have to understand it, which begins with observation. You'll be invited to engage in observation throughout this book, as you take a deeper dive into the hydrology and insect life of your space. Here in the beginning, you're encouraged to be a big-picture observer. There are few things to notice:

- How does the sun move across your property? How does this change during different parts of the year?
- Where does the water stream and pool?
- What plants are already growing here? What can the plants tell you about the features of that part of the yard? (For example, if moss is growing, you know that area stays damp and shady.)
- What makes various spaces suitable or not suitable for various types of plants? Are existing plants thriving on their own, or do they need resource-intensive maintenance?
- Do you regularly see animal tracks, pathways they've created through the grass, droppings, or other signs of activity?
- Do the existing structures, plants, landscaping features, and hardscaping elements create microclimates in the yard? (Read about microclimates later in this chapter.)
- What's the weather and climate like? (Read more about this in the coming section.)

As you start to envision your ecological garden, keep in mind these details. You are setting out to work with the conditions you are given.

CLIMATE CONSIDERATIONS

Plant types and varieties are each suited to certain climatic conditions. Expecting them to thrive in any other climate is setting up your garden for failure. While "typical" weather and climatic conditions are becoming less typical by the year, a knowledge of your area's weather and climate history is still the basis of where your ecological gardening decisions begin.

Temperature and precipitation. The most basic of climate data are the temperatures and precipitation that you can expect throughout the year. The National Weather Service is a good resource for finding your seasonal average temperatures and precipitation.

USDA Zone. The United States Department of Agriculture (USDA) tracks the average extreme low temperatures in every zip code in the country and charts them as Plant Hardiness Zones. From Zone 1a in the interior of Alaska to Zone 13b in coastal Hawaii, these zones are charted in 5°F (3°C) increments. USDA Zones (https://planthardiness.ars.usda.gov) are helpful in determining what perennial plants will survive the winter in your garden, and you'll find zone information on plant tags and in seed descriptions. USDA Zones are not the final word in perennial suitability, though. It's also important to know how long low temperatures will stay low and high temperatures will stay high, plus the kind of protection from the harshest elements your plants will find in your garden. By placing an emphasis on native plants in your garden design (you'll learn how in chapter 4), parsing through the USDA Zone details isn't necessary, as you already know the plant is suited to your climate.

Knowing which plants will grow successfully while providing necessary habitat for wildlife in your region begins with understanding the specifics of your climate and growing zone.

Snowpack. The snowpack is a means of protection for your plants. As snow covers the ground, it acts as an insulator, holding in the warmth of the earth and protecting plant roots from the harshest air temperatures. If you're in an area that sees consistent winter snowpack, even with a low USDA Zone, your plants' roots could be snug and safe under their white winter blanket, ready to create new life come springtime. The midwinter warmups and unpredictable snowfall we've seen in more recent years make snowpack less reliable, throwing some uncertainty into perennial plant selections.

Microclimates. While the USDA Zone covers each zip code, looking deeper within each zip code reveals microclimates, which are pockets of climatic features that differ from the climate around them. Your yard is full of microclimates that you can work with to design your garden. Both natural and man-made, here are a few microclimates that might factor into your garden design:

- *Heat sinks:* Building walls, fences, large rocks, and patios all absorb heat from the sun and reflect and release that heat into the soil and air immediately surrounding them. Southern and western exposures are particularly good at raising the temperature.
- *Windbreaks:* Wind dries out soil, and the strong air movement can be stressful for plants, especially as they're getting established. A fence, hedgerow, or other structure that blocks the wind creates a useful microclimate.
- *Low-lying areas:* Both water and cool air collect in low-lying areas. These are not necessarily negative microclimates, just conditions you need to be aware of.
- *Shade:* Shade cast by large plants, buildings, and other structures offers interesting spaces to construct shade gardens for plants that can't handle too much direct, harsh sunlight.

"Borrowed scenery" can be an important part of your garden's design, and it doesn't have to be as majestic as distant mountains to be valuable. A flowering tree across the street or a neighbor's decorative fence can become part of your own garden's design.

WORK WITH WHAT YOU HAVE

Incorporate into your garden plans anything existing on-site that has value, from a personal or ecological sense. Before you start tearing out plants or putting down pavers, mark and protect what you'd like to keep. This includes native plants and useful landscape features (swales or an existing privacy fence, for example). As you're dismantling and removing structures, stonework, or other hardscaping, consider what can be reused in your new garden plan. Think about each object's life span, from cradle to grave, in an ecological design.

What you have also includes the natural and built landscape around you. East Asian garden design uses the concept of "borrowed scenery," which is the idea that the background setting to your garden is also part of your garden. The orange tree hanging over the fence from your neighbor's yard, the large oak that casts shade from across the street, or the backdrop of a mountain in the distance are naturally part of your garden design. You don't have influence over whether these elements exist, especially those mountains, so the ecological choice is to accept and work their aesthetic into the space you're creating.

Whether you garden in a dry climate, like this gardener, or a mesic one, create a landscape that includes a diversity of plants with different shapes, sizes, and forms.

When installing paths, patios, and driveways, opt for permeable materials that allow rainwater to soak through.

PLANT CHOICE

While there are few definitive statements that can be made about ecological garden design—each garden will look different, after all—one constant is the importance of careful plant selection. In chapter 4, you'll read about matching plants' needs with your site's conditions. You'll also get to know the polyculture approach. A diversity of plants offers the most benefit to all of nature, and it has the bonus of providing the varying sizes, shapes, textures, and colors that good garden design requires.

HARDSCAPING

Hardscaping is the nonliving element of your garden space: walkways, seating, décor, and so on. You can choose to align these elements with your ecological principles, as well. A few ecological hardscaping ideas include:

- Instead of traditional concrete, look for permeable patio and walkway solutions. These allow rainwater to soak through them into the soil, whereas standard pavement and concrete cause water to run off.
- Use what you already have on hand. Shop at secondhand and salvage stores, search neighborhood for-free and for-sale listings, and otherwise reach for used items to avoid new construction and the resource use that comes with it. This is advice you'll read again and again in this book.
- Look for locally available materials and locally made products, rather than shipping something from afar.
- When you must purchase new, use items constructed from sustainably harvested and rapidly renewable sources, such as bamboo and willow, or that are made from a high percentage of recycled material.

Ecological GARDEN *Styles*

WHILE ECOLOGICAL GARDENING differs from what is considered permaculture, organic gardening, and sustainable gardening, there is plenty of overlap between concepts. Between them, you have no shortage of garden styles that can be considered ecological in focus.

RAIN GARDENS

Designed to capture water so it has time to soak into the ground, rain gardens can be any size and shape to fit your yard. The location, too, depends on your needs: You can place one in a wet, low-lying area, at the downspout to collect water from your roof, or along the downhill side of your driveway, where your yard might take on runoff from the surrounding pavement. You'll learn more about rain gardens in chapter 5, but know that these are beautiful and useful forms of ecological gardens in any area that sees regular rainfall.

XERISCAPING

A xeriscape (pronounced *zera-scape*) is a low-water landscape. Used in arid places, xeriscaping features drought-tolerant plants, mulches, and targeted irrigation only when needed. These ecological gardens are said to use 50 percent less water than traditional landscapes.

You can usually pick out a xeriscape design by its prominent use of succulents and native plants. They're typically well mulched, sometimes with gravel or sand, as keeping the soil covered in this way helps to retain moisture and lower the soil temperature.

You may think a succulent garden would be a boring green space, but just wait until the aloes and agaves bloom and the bees and hummingbirds show up to pollinate their flowers. You may be surprised by the diversity that cacti and succulents can bring.

Rain gardens designed to capture runoff and allow it to slowly soak into the soil are suitable for both large and small yards.

Low-water gardens are particularly useful in arid regions. They focus on drought-tolerant and native plants.

Prairie gardens are miniature versions of the grasslands that once covered the Midwestern United States. They are ideal for open, sunny yards.

If another goal of your ecological garden is to feed the humans living there, consider growing an edible landscape by incorporating edible plants into your garden's design. From fruit trees and shrubs to classic vegetables, create a layered mixture of annuals and perennials that also supports the ecosystem as a whole.

PRAIRIE MEADOW

Across the Midwest, we've lost native prairie to human development, including farmers' and ranchers' interests. There are movements to rehabilitate and bring back the prairie landscape on a large scale, and you can do that on a small scale in your own yard too. Native plants of all colors, sizes, and textures combine in diverse ecologically managed prairie areas. A meadowlike garden can do well in an open, sunny location. The native plants that evolved to grow in prairies and meadows typically prefer soils that are less nutrient-rich than traditional landscape plants.

These are less structured than what you probably think of as a garden, established by broadcasting or sowing seeds across the area—all species mixed together to create a dense, diverse planting of grasses and flowering plants. Meadow landscapes tend to look effortless and low maintenance, but getting them established takes time and care.

EDIBLE LANDSCAPING

Conventional, straight-row vegetable gardens with tilled soil, fertilizer applications, and pesticide use aren't exactly ecological in design. The row-crop design is meant for production and harvest efficiency and, on larger scales, ease of tractor use. Indiscriminate fertilizer and chemical pesticide applications are quick fixes to the slow work of building soil and creating a healthy growing environment.

Vegetable gardens like this also tend to be seasonal, leaving little food and habitat for insects and others during the off seasons. This is the type of gardening that centers on humans rather than holistically considering all of nature. None of this is meant to put down traditional vegetable gardens or their growers but is to say that a truly ecological approach accounts for your needs as the gardener, each plant's needs, and the needs of the rest of our ecosystem community members. Growing vegetables, fruits, and herbs in an ecological garden looks more like a landscape, grouping plants in communities. An ecological edible landscape uses a mix of annuals and perennials and prioritizes the needs of nature over human desires.

Many of the plants in an edible landscape provide more than one function. They can provide food for humans or other creatures, have medicinal properties, bloom at different times of the year, attract beneficial insects (which include pollinators), provide shade or ground cover, insert a pop of color or texture, and offer habitat.

Considering the toll taken by many modern agriculture and global food system practices, you could argue that growing even a little of your own food is a sound ecological choice. Here, you can opt out of the thousand-acre, chemically intensive monocrop system and the transport of food in refrigerated cargo holds over hundreds of miles or more. In your edible landscape, you may only grow a handful of fruits, vegetables, or herbs, but you can do so in a way that supports the ecosystem and your values.

Edible landscape maintenance. An important distinction between fruit- and vegetable-producing plants and those that are meant to be ornamental is the resources that the plants require. Resources include:

Nutrients: Many vegetables are known as "heavy feeders," meaning they will extract a lot of nutrients from the soil. Heavy feeders include tomatoes, peppers, squash, and melons. Other plants, in the legume or Fabaceae family, work with bacteria in the soil to deposit nitrogen for their fellow plants to use. Legumes include beans, peas, clover, peanuts, and astragalus. Including legumes in an edible landscape won't solve all of your soil-nutrition needs, but they're a good example of the benefits of a biologically diverse planting. An ecological garden that includes vegetables, and especially annual vegetables and nonnative edibles, may require regular soil testing and amendments to ensure proper nutrient levels. (Read about soil building in chapter 6.)

Water: Similar to a more specific nutrient requirement, food-producing plants need more regular water too. The standard advice for vegetable growers is 1 inch (2.5 cm) of water per week. This is where grouping plants into communities with similar maintenance requirements is important, so you can spot-water the garden and not waste water on the plants that don't need it.

Time: Already you can tell that with the soil and water requirements, edible landscapes require more of a time investment than other types of gardening. In addition to the regular upkeep, you have to replant annual plants, remove spent plant material, and keep plants around them pruned so they have the proper amount of space necessary to thrive.

Climate: Thoughtful consideration for the types and varieties of fruits, vegetables, and herbs you add to your landscape includes their climate requirements. For perennial plants, look at the USDA Zone as well as your local weather conditions. For annual plants, consider their needs (frost and heat tolerance, specifically) compared to your climate. Investigate season-extension options, such as a floating row cover to add a few degrees of temperature protection for frost-sensitive plants.

Native edible plants, such as this serviceberry, provide pollen and nectar to native bees when in bloom and then berries to both birds and humans later in the season.

Access: When growing edible plants, you need to be able to get into the garden to harvest them. Plan pathways carefully, and realize you may need to move or prune plants, especially perennial plants, as they grow and crowd your access to their neighbors.

Support systems: Vining plants and those that get bushy and heavy as they put on fruit do best with some form of support. There are many beautiful vining plants that don't produce food that also need support, but the right support becomes especially important for food plants. Plants laying on the ground are more susceptible to pests and diseases, the fruit will pick up dirt if it's been lying on the ground, and you have to do a lot of bending and lifting to harvest vegetables from ground level, as opposed to from a support system. Whether you install supports in the form of trellises and cages, or you take the time to train these plants to grow along trees and other structures in the garden, they may need your help to grow vertically for their best health and production.

Sharing with wild neighbors: Thinking back to the conventional garden, great care is taken to exclude birds and wildlife from its borders, as the food grown within is for people. If you're including edible plants in your ecological garden, it stands to reason that you plan to eat the fruits of your labor. You may need to decide how much you're willing to share with your wild neighbors. This is part of determining your own ecological gardening standards. You could grow enough to go around, or you may rather maintain a section fenced from the rest so you're sure to have a harvest for yourself.

Alternative edible plants. In considering adding edible plants to your landscape, you aren't limited to typical vegetables like tomatoes, pumpkins, and corn. Think about edible flowers, including cascading nasturtiums, delicate jewels of opar, and cheery calendula. Even America's native sunflowers are edible. Herbs of all kinds are edible, many are medicinal, and they can add the sensory experience of an uplifting scent to your ecological garden. Many fruits, too, are native to where you live, often perennial, and as interesting to look at as they are delicious to eat. This includes fruit trees, as well as bushes and vines.

Food forests are layered food-growing gardens that mimic the layers found in a natural forest from trees down to ground covers and edible roots. They are lower maintenance and more ecologically sound than traditional row-based vegetable gardens.

Around the world, the biodiversity of food crops is being lost to consolidation and commercialization of the food system, war, and climate catastrophes. In your own ecological garden, you have the opportunity to boost biodiversity and plant rare, heirloom, and otherwise unusual edible plants. Search for suppliers of open-pollinated and culturally important seeds, and you may be surprised at the wealth and diversity of crops you've never heard of. Among your neighbors and American Indigenous seed keepers, you may even find crops that are native to your area.

FOOD FOREST

Mimicking forests in nature, a food forest is a layered approach to an edible ecological garden. Now most commonly considered a modern permaculture technique, the roots of food forests lie with Indigenous cultures. A food forest has seven layers: canopy trees, understory trees, shrubs, ground-level plants, ground covers, roots (grown for food or as soil stabilizers), and climbing plants. Each plant in a food forest is intentionally chosen for its role in the landscape. This landscape is meant to be low maintenance and self-sustaining. Most of the plants are perennials, and others are annuals or biennials that spread or reseed on their own.

While the word *forest* conjures thoughts of multiacre plantings, you can put the concept to work in a space as small as an urban backyard. Here, instead of starting with a towering oak canopy layer, you might let more reasonably sized apples and figs, which are typically understory plants, act as your canopy, and build the rest of the layers below them.

From habitat and food to soil retention and noise blocking, hedgerows provide multiple benefits for humans and nonhumans alike.

HEDGEROWS

Long rows of trees, shrubs, and grasses have been used for centuries in Europe as a tool for landowners to delineate property lines. From an ecosystem services standpoint, these dense plantings of mostly perennial plants do much more.

Hedgerow benefits include:

- **A visual screen:** Densely planted, this living fence line is a privacy screen for your property, as well as a backdrop to set other garden spaces against.
- **Sound buffering:** A hedgerow breaks up the way sound travels across the neighborhood.
- **A wildlife corridor:** Rabbits and other wildlife will use this cluster of plants as safe cover to travel across otherwise open space.
- **Habitat:** Insects, birds, and wildlife can make homes in hedgerows.
- **A food source:** Those same insects, birds, and wildlife can forage here, too, depending on the type of plants you choose. These visitors include pollinators and beneficial insects.
- **Fruits and nuts for you:** The wildlife aren't the only creatures who can enjoy hedgerows' edible benefits. Choose plants that produce what you also would like to eat.
- **A windbreak:** Windbreaks offer protection from erosion and water evaporation, plus they make the outdoors a more pleasant place to spend time. A well-designed hedgerow can reduce wind speed by up to 75 percent.
- **A source of shade:** In a hot climate or a yard with little shade otherwise, a properly oriented hedgerow can block harsh rays.

- **Soil building:** Perennial roots improve the soil structure and habitat for soil-dwelling organisms, creating an area rich in organic matter.

- **Soil retention:** Deep perennial roots hold soil in place on slopes, which is becoming more important as heavy rainfall makes hillsides less stable.

- **Year-round visual interest:** Whether evergreen or deciduous, a living hedgerow is a more interesting feature than a plain privacy fence in any season.

Because they have such value in the landscape, hedgerows are encouraged and incentivized by conservation farming programs. The larger the hedgerow, the greater the benefits, but home gardeners can take notice and incorporate them into ecological landscaping plans of any size.

Hedgerow composition. You might think of a hedgerow as a tall, uniform, evergreen fence and wonder how it can be a source of biodiversity. You might also think that something that formal doesn't fit your gardening style. Hedgerows can be designed as formal, managed hedges. They also can be a diverse composition of multiple types of plants that provide the most benefits to your community.

Browse the list of hedgerow benefits above and pick out those that are most important to you. Then choose plants that fit those purposes (as well as your climate and planting conditions). Ideally, you'll have plants that are blooming at different times of the year and at least some that maintain their shape and fullness year-round. Fill in the space between taller trees with shorter shrubs and the space between shrubs with grasses and flowers to create a picture of colors, shapes, and textures.

Hedgerows consist of a collection of densely planted plants from multiple layers. In this illustrated example, small trees, shrubs, evergreens, perennials, and ground covers combine to create a biodiversity-supporting system.

HEDGEROW PLANT CHOICES

There's almost no wrong choice for a hedgerow plant, as long as the plant is not invasive in your area and its needs match your climate and site characteristics, but it's best to seek out species that are native to your region and will grow there with minimal care. Many potential hedgerow plants offer multiple benefits. This chart offers just a few of hundreds of multipurpose plant options. All of the plants here are perennial, so they're soil builders and stabilizers, and they're meant to be visual, sound, and wind barriers.

HEDGEROW PLANT	WILDLIFE HABITAT	WILDLIFE FOOD SOURCE	HUMAN FOOD SOURCE	ATTRACTS POLLINATORS	YEAR-ROUND VISUAL INTEREST
American hazelnut (*Corylus americana*)	X	X	X		
Dwarf serviceberry (*Amelanchier spicata*)	X	X	X	X	
New Jersey tea (*Ceanothus americanus*)	X	X	X	X	
Northern wild raisin (*Viburnum cassinoides*)	X	X	X	X	X
Oregon grape (*Mahonia aquifolium*)	X	X	X	X	X
Pagoda dogwood (*Cornus alternifolia*)	X	X		X	X
Salal (*Gaultheria shallon*)	X	X	X	X	X
Shrubby St. John's wort (*Hypericum prolificum*)	X	X		X	
Silver bush lupine (*Lupinus albifrons*)				X	X
Strawberry tree (*Arbutus unedo*)	X	X	X	X	X
Summersweet clethra (*Clethra alnifolia*)		X		X	X
Sweet fern (*Comptonia peregrina*)	X	X	X	X	
Sword fern (*Polystichum munitum*)	X				X
Tobacco brush (*Ceanothus velutinus*)	X	X		X	X

Green roofs are a great way to bring ecological gardening to urban environments and provide much-needed resources for insects and birds.

GREEN ROOF

A green roof is not a garden in the traditional sense, but it is a form of ecological gardening. The rooftop is covered in a shallow growing medium and planted with suitable plants. This planting can be part of a home water-collection system. It also provides stormwater management by absorbing some of the rainwater and slowing the rainwater's path to sewage and septic systems. While installation costs are substantial, at $15 to $20 per square foot (929 square cm), the potential energy savings are also substantial. A one-story home can save 20 to 30 percent on cooling costs, thanks to the insulative properties of the green roof.

HÜGELKULTUR

Looking to the ecological garden as a source of water management, *hügelkultur* checks that box and others. In areas with plentiful precipitation, these German-designed mounded beds can direct, absorb, and retain water while providing healthy soil for the plants rooted in them. A *hügelkultur* bed starts with a dug-out trench that's then mounded with organic materials, including large logs that will take a long time to break down, and smaller pieces, such as leaves and twigs, that will decompose more quickly. The soil that was removed is mounded on top to create a berm you can plant into.

A *hügelkultur* bed helps manage water in several ways:

- Properly placed, the berm catches water as it's moving across the ground and gives the water more surface area to soak into. (You'll read about berms and swales in chapter 5.)
- The organic material inside the berm acts as a sponge, soaking up as much water as it can. This water is released as the soil dries, and plant roots have continual moisture.
- At the same time the organic materials are absorbing water inside the berm, they're being broken down by soil organisms. The work of these organisms builds soil organic matter, which improves the soil's ability to hold and filter water.

All this organic material feeding the soil means *hügelkultur* beds are generally self-sufficient in their nutrient production, not requiring future soil amendments. Additionally, a *hügelkultur* bed can provide a wind block and adds depth and lines to your garden design.

A completed hügelkultur mound ready for planting.

Your Design Team

Is an ecological garden design the sort of thing you should do on your own? It can be, but it doesn't have to be.

A professional landscape designer can present you with a garden plan tailored to your climate and your personal ecological garden principles. This person most likely studied horticulture or landscape design and understands the nuances of these subjects, along with the rules of garden design. It may be worth exploring work with a garden designer, if that makes sense in a cost-benefit analysis for you.

For design-it-yourself gardeners, there's no shortage of resources out there. Apart from the obvious books, websites, and videos, one resource may be even more valuable: other gardens. Garden clubs host local garden tours, universities have demonstration gardens, and botanical gardens exist across the country.

On an even smaller scale, there are gardens in yards all around you. Most people love the chance to talk about their gardens and plants, and you can learn so much about design choices just by striking up a conversation.

It's also nice to get feedback on a plan from someone who has experience with ecological gardening. Look to your network of gardener friends whose places you admire, and don't forget about your local Master Gardeners. Managed by the Cooperative Extension network, an important component of the Master Gardener program is volunteering to help gardeners in the community. A Master Gardener may be available to talk about garden design and ecological garden concepts with you. (Mind you, not all Master Gardeners will be taking an ecological approach, so be sure you know where they stand before you get started.)

LOSE *the* LAWN

THE GREAT GREEN LAWN has been called America's largest irrigated crop. Covering 62,500 square miles (162,000 square km) of ground, our lawns span an area more than thirty-one times the size of Delaware (about four times the size of Switzerland).

It's thought that manicured lawns became a status symbol in Europe in the 1700s, signifying a property holder had enough land to dedicate some of it to a frivolous, rather than food-producing, crop. By the 1950s, new lawnmower equipment, turfgrass varieties, and pesticides and fertilizers provided by turfgrass companies made lawns the norm in the United States.

Lawns have plenty of positives: They're uniform to maintain; they can stand up to foot traffic, which is great for yards with kids playing in them; and thousands of plants compose a lawn, and each of them filter air, sequester carbon, and help maintain soil structure.

It's a lawn's negative attributes that are concerning to ecological gardening. These include:

- Lawns are essentially monocultures. There are different types of grasses, of course, but by and large, most plants in a lawn are grasses that all go dormant at the same time. They offer little in the way of forage value to insects, birds, and wildlife, because they don't often get to go to seed or flower before they're mowed. They also all look the same, which is not very interesting.

- Maintenance is a time- and energy-dependent task. Most turfgrasses require regular mowing, which requires the use of power equipment (or reel mowers, which are notoriously difficult to operate under all but the most ideal conditions). With many lawns being made up of plants ill-suited for the climate, they need regular watering to thrive.

Who needs a back lawn when a back garden is so much more inviting and resilient!

- The $30-billion-per-year lawn-care industry has convinced homeowners their lawns require herbicides and pesticides. These chemicals disrupt and destroy insect populations and soil microbial life, and their runoff pollutes waterways and groundwater. They could also be harmful to the humans and animals that live, work, and play around lawns.
- Turfgrass needs full sun and plenty of water to look its best, which are hard to come by in many areas. These thirsty plants compete with others for water and nutrient resources.
- Lawns take up space that could be used for more interesting and ecological purposes. You only have so much land available to steward, and you can interact with it in more creative ways.

TRANSITION INCENTIVES

Seeing value in the benefits of using residential outdoor spaces for plantings other than turfgrass, some municipalities offer incentives for homeowners to replace a water- and chemical-intensive lawn with a more ecological choice. Several conservation agencies and water districts in California, for example, offer rebates for installing rain gardens and removing turfgrass lawns.

LAWN ALTERNATIVES

Going from full-conventional lawn to no lawn at all is a big adjustment—for homeowners and for neighbors. Tearing out your whole yard to put in a garden may not be right for you, and in fact, starting a garden on a small scale is a great idea. A garden space of just 30 feet (9 m) by 3 feet (1 m) equals 90 square feet (8 square m) of an ecological oasis in what was once standard green grass.

Lawns replaced with a collection of regional native plants offer stunning and ecologically vibrant results.

There are a few ways you can make progress toward a more ecologically managed outdoor space while maintaining some of what's familiar to you:

Mow less. While this is the simplest alternative-lawn option, it may also be the most contentious. Regardless of what those around you might think, it is possible to just not mow your turfgrass as often as prescribed by the lawn-care industry. There are studies showing bees, butterflies, grasshoppers, and other insects benefit from this practice. One study in Massachusetts found that lawns mowed every three weeks had two-and-a-half times more lawn flowers, such as dandelions and clover, than lawns mowed every week or so, while lawns mowed every two weeks had the greatest number of bees.

Have less lawn. Maybe you like having a lawn because you enjoy the way the grass feels under your bare feet and you like to have a place for your dog to run. Also, maybe you're under regulations that require a certain amount of your yard be actual lawn. It's possible to convert part of your yard into an ecological garden while part of it is turfgrass. This doesn't need to be an either-or proposition. Look back at your personal ecological gardening purpose to decide what's best for you.

Use alternative grasslike plants. Turfgrass varieties are the traditional lawn plants, but other grasses and ground covers can create a lawnlike look to meet landscaping regulations while requiring fewer resources to maintain.

Choosing the right alternative-lawn plants for your yard involves the same thought process as choosing the right garden plants. Look for those suited to your climate and yard conditions.

Some turfgrass alternatives include:

White clover: Though it is not native to North America, white clover is hardy from north to south across the United States. White clover is a low-growing, drought-tolerant plant that can support certain bee species. Establishing clover in a lawn takes some work, as you have to remove the existing lawn plants, heavily seed the area, and keep it well watered until the clover becomes established. It doesn't require mowing, unlike a traditional lawn. As a leguminous plant, clover works with soil life to deposit nitrogen in the soil for other plants to use. Clover does die back in the winter, leaving the yard looking brown come cold weather.

Sedges: Particularly suited for wetter climates, several sedge varieties can be used as low-mow lawn plants. Look into California meadow sedge, Appalachian sedge, and Caitlin sedge. Sedges grow in clumps, rather than in turfgrass's spreading fashion. As they mature and grow together, they cover the ground. This is a different sensation for walking on, but it's visually appealing. Sedges prefer shade and will tolerate drought there; in the sun, they might need watering. They need mowing just a few times a year.

Buffalograss: Native to the central United States, buffalograss varieties grow to just 4 to 8 inches (10 to 20 cm) tall, meaning it needs little mowing. It does best in full sun and does not tolerate too much water or soggy soils. Buffalograss goes dormant in cold weather.

Alternative grasslike plants and ground covers offer a solution for smaller urban yards where a lawnlike appearance is desired without the need to mow.

No-mow mixes: Seed companies have produced their own mixes of turfgrass-alternative plants for low-maintenance lawns in various climates. If you're reaching for one of these, be sure you understand what each species in the mix looks like, the care it requires, and how it functions. Some no-mow mixes are primarily grasses, so be sure you know what you're buying.

Ground covers: If you're less concerned about your yard having a grasslike appearance, durable ground covers can be employed here, with a particular focus on those that are native to your region. Ground covers give you the flexibility to have outdoor lawn space without the typical maintenance and upkeep of a lawn, and ground covers are often flowering plants, adding to the ecological value of your landscape. Planting ground covers is a nice option when you're tired of the lawn but not ready to dedicate your whole outdoor space to a managed garden. These are also a nice option for particularly steep yards, where the slope makes mowing or gardening a risky task.

LAWN TRANSFORMATION IN REALITY

Whether your lawn-reduction end goal is a yard that's more resilient, provides ecosystem services, or is less work to maintain, expect that it will take some time to get there. In addition to the work required to do away with your existing turfgrass, the first few seasons with your new lawn require upkeep. All plants need water to get established, and when you introduce the new plant species, whether by seed or plug (small plant), other seeds and rhizomes will be making their way into your yard. It's essential to stay on top of weeding and watering so your newly introduced lawn plants can take root, spread, and thrive. Once the plants' root systems are established, they'll minimize weed pressure over time.

Sedges make an excellent lawn alternative.

Finding Middle Ground

To appease the manicured-lawn crowd and meet landscaping ordinances, you have options for a middle-of-the-road approach to your ecological gardening objectives.

Be intentional. Part of what makes people think nonconformist gardens look messy is that the gardens are installed without a plan. When you are intentional about your plant choices, borders, hardscaping, and so on, your garden will have a natural flow and progression. This gives the impression that your neighbors can trust that you know what you're doing, because you do.

Delineate borders. A mowed edge, a fence, a low shrub line, or another setback between the garden and the sidewalk or property line are sometimes outlined in landscaping ordinances. Even if not specifically requested, this border gives peace of mind to neighbors and passersby that your naturally managed garden isn't going to spill over onto their yard or right of way. Borders also help to make garden spaces look more orderly and provide a line for the eye to follow. Sometimes called a *cue to care,* adding a border or mowed edge is an easy way to signal to neighbors that an area is not neglected.

Mowed edges and paths offer "cues to care" that let neighbors know your garden is well tended and intentional.

Go slowly. While this isn't meant to be a psychology book, it's worth noting that people don't like change. A gradual progression from lawn to ecological garden can help neighbors get used to the idea of a naturally managed space taking shape in your yard. Starting small will also help you to understand the time and resources your garden requires so you don't overcommit.

Be thoughtful in your choices. You're here to make a great first impression.

Leave *some* leaves. The "leave the leaves" concept is one that tends to strike a nerve with neighbors. Regardless of the importance of the insect habitat that fallen leaves provide and the nutrients that decomposing leaves return to the soil, the practice of letting leaves lie on the ground is often misunderstood. Neighbors might complain that the leaves make your yard look messy or will kill the grass underneath. They probably have concerns about your leaves blowing onto their property, which is understandable, considering how much time they spend picking up their own leaves. Living in a community with other people requires compromise, and the leaf issue is one area where you have flexibility.

Consider these ideas:

- Allow leaves to sit on certain areas, perhaps areas less visible to your neighbors, while keeping other parts of your yard clean.
- Use hedgerows and other dense plantings to both create a visual screen and corral the leaves in your space so the wind doesn't blow them around the block.
- Collect your leaves in your compost bin or make leaf mold. "Leaf mold" is a term for partially decomposed leaves that are rich in nutrients and make a great natural mulch. This collection of leaves will provide the habitat that insects need to overwinter and preserve nutrients for you to return to your landscape while also creating a tidier appearance for your yard.

Stand up for your stems. While the dead plants that you leave standing provide nesting spots for beneficial insects, forage for birds, and visual interest in the winter garden, some neighbors just see dead plants.

Here, too, there are some ways you can compromise:

- Be thoughtful in where you locate your plantings. Maybe plants with pithy stems don't take center stage in the front yard; there could be a less visible corner for them elsewhere.
- Separate your "wild" plantings from the managed community with hedgerows.
- Trim the plants somewhat but not all the way to the ground (you'll learn more about this technique—sometimes called the *Chelsea chop*—later in this book). This may be effective in areas with plant-height ordinances.

Be prepared to educate neighbors and visitors about your garden and the creatures living in it. A "bee house" might sound scary to someone afraid of bees, but once they understand the docile native bees that call such a structure home, they'll be less likely to push back.

"No mow" or "low mow." No-mow May—or March, depending on where you live—encourages property managers to forego mowing the lawn for a month in early spring, when native pollinators are coming out of dormancy and foraging on early blooming flowers. These campaigns have raised awareness about the role that our lawns can play in the health of the neighborhood ecosystem and beyond. Not all communities appreciate this effort as much as the bees do. If you live in one of those communities, you could be better off picking your battles and supporting your resident pollinators in other ways, including:

- Low-mow spring asks you to mow less often so flowers have a chance to bloom and bees can feed on them for at least a few days before you knock them down. While you might not be able to get away with a whole month without mowing, you could extend the time between mowing.
- Be sure your landscape includes flowering plants elsewhere, even if that means putting some annuals in a container garden until your landscape perennials start blooming.
- Tuck an unobtrusive pollinator house (or several) into your landscape.
- Reduce the size of your lawn so mowing becomes less of an issue anyway.

Educate. Yes, educating your community about your ecological gardening practices was already mentioned as a means of being neighborly, but this is important enough that it's worth mentioning again. The more people understand what you're doing in your ecological garden and how it's benefiting everyone, the less pushback you're likely to receive.

LEARNING ECOLOGICAL *Gardening Practices*

THIS CHAPTER MENTIONS several techniques used in the garden, such as composting, managing fallen leaves, and managing past-season plant materials. You'll read more about all these and other gardening techniques in chapter 6.

Caring for an ecologically vibrant garden requires a shift in mindset from the methods and techniques used to tend a more traditional garden. The shift can be a challenging one for some gardeners, and that's why we dedicate an entire chapter to the subject later in this book.

FIRE-WISE GARDENS

WILDFIRE IS A TOPIC for ecological gardening because it is an essential and inescapable part of our natural landscape. It is possible to have a beautiful garden that contributes to your home's and neighborhood's fire resilience. This requires understanding basic fire behavior and how your landscaping interplays with it.

THE NATURE OF WILDFIRES

In mainstream society, wildfires are maligned and feared. There's good reason for this. Wildfires cause loss of life and property; create related environmental hazards, such as landslides and water pollution; increase insurance rates; and disrupt whole communities. While wildfires can be destructive to life as we know it, they are natural systems. It's up to us to learn how to best protect ourselves from them.

In ecological, Indigenous, and traditional land management, wildfires are understood and respected as a natural renewal process. Fire is an important tool to manage wildlife and either suppress or encourage plant growth.

Wildfire safety is a more relevant topic now than ever before. The majority of counties west of the Mississippi River are at a relatively moderate to relatively high wildfire risk, according to Federal Emergency Management Agency National Risk Index. Between 2000 and 2021, 70,600 wildfires burned 7 million acres (18 million square km) on average each year, whereas in the 1990s, we saw a higher number of wildfires, but half the acreage burned. While the largest and most destructive wildfires tend to be in the western United States, most wildfires, in fact, burn in the eastern half of the United States.

This miniature model of fire-wise landscaping at the San Diego Botanic Garden provides a visual example of how to landscape in wildfire-prone regions.

PRINCIPLES OF FIRE-WISE LANDSCAPING

This is not meant to be an all-encompassing guide to creating a fireproof yard. (There is no such thing as a fireproof yard.) The ideas here will help you make decisions to turn your landscape into more of an asset than a hazard in fire management. If fire safety is of particular concern for your area, use these ideas as a starting point, but do additional research that includes gathering information from your local fire authorities.

Some of these basics will contradict the principles of ecological gardening that you read elsewhere. For example, suggesting that you remove leaves and dried plant matter from your yard because they're flammable is at odds with the idea of leaving plant material standing for the insects, birds, and other wildlife. There is nuance here. Particularly within your home's defensible space, this combustible material poses a greater risk of fire than it does a reward to wildlife. This is your opportunity to look back at your personal ecological gardening values and decide what is best for your situation.

Zone 0
Zone 1
Zone 2
Zone 3

Fire-wise basics include:

Create defensible space. The combustion-free buffer around your home is called *defensible space.* If a fire is approaching, it could slow or stop at this dividing line because you've separated your home from the flammable materials surrounding it.

Defensible space can be visualized in four zones:

- **Zone 0** defensible space is the area within 5 feet (1.5 m) of structures and foundations. Keep this area clear of anything with fire-ignition potential, including woody and dried plants, landscaping features, tools, outdoor furniture, and items stored beneath your deck. Gravel, pavers, and concrete are nonflammable landscaping features appropriate for this space. You can also use well-maintained succulent plants, which are rich in moisture, in this area.
- **Zone 1** defensible space extends from 5 to 30 feet (1.5 to 9 m) from your home. Continue to keep this area clear of flammable vegetation. Propane tanks should be clear of vegetation and combustibles by at least 10 feet (3 m) on all sides.
- **Zone 2** defensible space is the area from 30 to 100 feet (9 to 30 m) away from your home. Shrubs and trees are appropriate here, but put space between them to reduce the fire's ability to jump from one to another. Firewood, propane tanks, and other highly combustible materials need to have a 10-foot (3 m) buffer from flammable vegetation and materials.
- **Zone 3**, from 100 feet (30 m) and beyond, is more the purview of community, forest, and grassland management than it is that of ecological gardening, and should be part of a larger community fire-management plan.

While these guidelines simplify the defensible-space concept, your local fire marshal may have different rules for your neighborhood. Always follow their guidance.

Maintain your landscape. While the point of an ecological garden is to not need water resources to maintain a landscape, in times of drought and fire risk, watering plants to support their health may be necessary. Dried plant materials are prime fire fodder.

Trim tree branches to at least 10 feet (3 m) from your roof and chimney to prevent flames from having direct access to the structure. Monitor trees and shrubs for health, and remove dead or diseased woody material each fall and winter. Pick up leaves and sticks in accordance with your defensible zones.

Understand fire's movement. Moving uphill, fire can spread rapidly because the flames' vertical reach creates a ladder effect from plant to plant and especially from tree to tree. Increase your defensible space and the space between trees, shrubs, and flammable objects as the incline of your property increases.

Use fireproof or fire-resistant materials. Wood is an obvious fire fuel. When planning an ecological landscape in a fire-prone area, consider what wood items you can replace with less-combustible materials. Some states, including California, even have deck-board materials standards. Check with your state and local fire marshal.

Choose plants carefully. Native plants, suited to your climate, are more likely to withstand the weather that's in store for them. Healthy plants, full of moisture, are less flammable than stressed, dried-out plants. Because maintenance is essential to fire safety, choose plants that require a level of upkeep that matches your ability and ambition. Choose succulents and other plants that store moisture in their leaves and stems and don't produce a lot of dry, woody, fine debris. Plants that are high in volatile oils (for example, pines and conifers) may be more likely to ignite. Seek advice from regional fire-prevention experts and your local Extension office for plant lists appropriate to your specific part of the country.

Responsibly dispose of plant materials. You are responsible for the plant materials that come from your property. A moist, active compost pile is a good means of disposal. Plant materials disposed of in unmanaged piles and on public land can add to a fire's fuel.

The idea of being a good neighbor is a building block to the rest of ecological gardening. Combining your aesthetic sensibilities with nature's requirements and measuring all of that against your human community's rules and norms is a balancing act. It's also very possible, with the right approach. The benefits that ecological gardening offers are worth the effort.

Now that you're thinking about a garden space from a perspective broader than simply what it can do *for you,* it's time to learn about the life to be found in your garden, above ground and below, visible and microscopic.

Hedgerows in the Fire Zone

You may have become sold on the idea of a hedgerow after reading their long list of benefits earlier in this chapter. Hedgerows can also be a danger in a fire-prone area. Fire experts stress the importance of keeping hedgerows away from structures, as they're a source of fuel. If you were to opt for a hedgerow in an area with a wildfire risk, choose high-moisture plants, which are less flammable, and keep them pruned, regularly removing dead and dry material.

Combining your own sense of aesthetics with the needs of nature can indeed yield beautiful results.

CHAPTER 3

YOUR *Garden* HABITAT

THE GARDENS AND LANDSCAPES we build and maintain around us serve many purposes. They provide us with food to fuel bodies and beauty to soothe our souls. They are creative art projects and spaces to gather with friends and family. They filter stormwater and cast cooling shade on hot summer days. But sometimes we forget that the landscapes around us are also habitat for countless other living things.

Often, we think of things like wildlife habitats and ecosystems as existing apart from our daily lives in places like national parks and wildlife preserves. It is easy to imagine nature as being something separate from us, something that happens somewhere else.

The reality is that we exist in and depend on the same ecosystems as every other living thing, and we all walk through wildlife habitat every day when we head out the front door to work. This is true whether or not you have planted up your yard with native plants and have a little sign out front that says "Pollinator Garden." Even a very traditional landscape in a generic neighborhood is teeming with life.

To start to understand this, grab a chair, a notepad, and a pen. Take all that outside and have a seat. Then, pen in hand, try to list every animal—be it a bird, insect, or mammal—you can see or hear. Don't worry about identifying everything, just scribble it all down. The point here is to get a sense of the diverse world around you, so whether you write down "small brown bird" or "sparrow" doesn't matter.

You'll probably start with the big, obvious things. Your neighbor and their dog going for a walk. The squirrel burying acorns. The crows having an argument in a tree up the street. Oh, yeah, that mosquito that just bit you? Write that down too. Living in nature is beautiful, but it is also sometimes itchy and annoying.

As you keep observing, you'll see smaller and smaller things. A little white butterfly—or is it a moth?—fluttering by. A bumblebee pushing into a flower. Get down on the ground and look even more closely. You'll almost certainly see an ant—probably several ants of multiple different species. Peer closely between the grass or poke around in the mulch and you'll find beetles, centipedes, worms, and isopods.

If you grabbed a shovel and turned over a bit of soil, you'd find even more things—many of which you probably don't have a name for—running around. If you grabbed a microscope you'd be stunned by the sheer number of tiny little living things busily going about their lives on a tiny patch of ground you've probably never thought much about.

Not all garden residents are adored by every gardener, but they all do play a role in a balanced garden ecosystem.

In just a few moments, you'll probably be able to list dozens of different animals living all around you. How many, do you think, would you find if you really took the time to pore over every detail, look under every rock, and identify every single little wriggling thing you saw? Well, in 2020, while in lockdown for COVID-19, three scientists decided to try just that. For a full year, they tried to document every single species they could find on the lot where they lived in a suburb of Brisbane, Australia. This was a small, urban property, just shy of one-tenth of an acre (405 square m), with nearly half that space taken up by their home. Going into the project, they and their colleagues guessed that they might find a couple hundred species, but their actual results shocked them: After a year they had recorded fully 1,150 different species. This was no nature preserve; there was nothing special about the land. It was just a generic, small urban property, bursting at the seams with more life than anyone would have guessed.

The same is true of your landscape, wherever you are, or however large or small it is. The land you live on is rich, diverse wildlife habitat. And with some care, you can learn more about it and make it even more vibrant and vital.

Sitting outside and listing all the living things you find in your garden is a great first step to really understanding and celebrating the diverse habitat that is your landscape. And learning even more about the birds, insects, mammals, reptiles, and amphibians that call your garden home is a wonderful way to get committed to creating an ecologically vibrant landscape. Learning and observing may not seem like ecologically important things to do, but they can really make all the difference.

The first reason to learn more about the life in your garden is for the sheer joy and beauty it can bring to your life. Slowing down, you'll suddenly discover that there are far more beautiful songbirds and brightly colored butterflies in your garden than you ever imagined. Once you take a closer look, you might discover that moths you've always overlooked are actually beautifully patterned, and that even lowly beetles often have colorful, iridescent shells. If we don't take the time to look, we only notice the other living things around us when they are a problem: mosquitoes that bite us or ants that get in the house. But only a tiny percentage of insects ever bite or otherwise bother humans. Most are just living beautiful, intricate lives waiting to be discovered.

The next benefit of learning more about the life in your garden is the way it will transform gardening with an ecological point of view from a chore into a pleasure. It is easy to feel like gardening—and living—with an eye to the environment around us as something that we *should* do, something that is annoying or difficult, but that we are morally obligated to do. When we do something out of a sense of moral obligation it is rarely fun or pleasurable. We may do it, but we're not necessarily happy about it. But when doing the ecologically sound thing changes from an abstract concept to a concrete reality, suddenly it stops being something you *should* do and changes into something you *want* to do.

When you walk out the back door one day and half a dozen beautiful butterflies swoop around you, you aren't going to want to harm them in any way. When you spend a summer evening watching clouds of glowing fireflies bobbing over your yard—and realize none of your neighbors have nearly as many—you'll be delighted to leave some places for them to overwinter during your fall garden cleanup. And when you spend a few minutes every morning watching a mother bird feed caterpillars to her babies in the nest she built in a shrub by your front door, suddenly instead of worrying about holes in your leaves when you see caterpillars on your plants, you'll be delighted that your garden is providing plenty of food for the next generation of songbirds.

The final reason to learn more about the living things your garden supports is what you can share with your neighbors. As mentioned in the last chapter, reducing your lawn, watering a little less, or letting more leaves and stems naturally decay in place rather than cleaning everything up doesn't always go over well in communities where manicured lawns and neatly trimmed shrubs are the norm.

It's easy to appreciate an eastern black swallowtail butterfly sipping nectar from summer flowers, but it may be harder to appreciate its larva eating the leaves of your golden Alexanders plant (*Zizia aurea*). But golden Alexanders evolved side by side with this herbivore. The plant and the caterpillar (and the bird who may eat it) are all parts of a functional ecosystem, each with its own invaluable role.

Just saying it is good for the environment isn't usually very convincing. Pointing out a specific butterfly and explaining why it needs these conditions to thrive makes the case a lot more effectively. Share the specific beauty and joy you find in your garden and you just might convince others to change their gardening practices as well. And who knows, they might then tell their neighbors, and you will have started a chain reaction capable of spreading through your community and transforming not just your yard but your whole neighborhood.

LEARNING

LEARNING ABOUT THE OTHER living things in your garden is central to loving and supporting them. But how do you go about doing that? There are a lot of different ways to approach it.

SIT IN THE GARDEN

The first and simplest way to start learning about what you share your garden with is to do the exercise outlined in the beginning of this chapter. Take a little time every now and then and just sit outside and observe. This can be a wonderfully meditative experience, really connecting you to the greater world. Beyond just watching the life that surrounds you, take the time to get a closer look at the lives happening in your garden. Maybe that means finding a safe distance to watch a mother bird taking food to the babies in her nest or taking the time to watch the lifecycle of an unfamiliar insect unfold.

Spend time in your garden observing the life found within it. Look closely and take your time. Nature's pace is slower and deserving of our attention.

Rearing Caterpillars

One simple way to get a better view of the insects in your garden is to raise a caterpillar through all of its life stages. When you find a caterpillar munching on a leaf, bring it inside into a safe enclosure and provide it with food (more of the same leaves from the plant it was found on), watch it grow and expand, and eventually observe it forming a chrysalis or cocoon before finally emerging as an adult butterfly or moth. Children—with supervision—will love this activity, but that doesn't mean you can't enjoy and learn a lot from it as an adult as well. Keep in mind that the practice of rearing certain species of butterfly inside can prevent them from migrating, and caterpillars reared in captivity may be more prone to certain parasites and diseases. In general, caterpillars should be allowed to live outside in the natural world. However, it isn't a harmful practice to rear a single caterpillar to get an up-close look at the process. Here's how to do it.

Though they are best left outdoors in the natural world, rearing monarch caterpillars indoors on milkweed stems offers a simple way to get a closer look at the wonders of nature. This one-time experiment can result in a lifelong appreciation for nature's cycles for kids and adults alike.

1 **Identify the caterpillar, if possible.** This is an important step because it will help you care for it properly. If you can't figure out the exact species, raising it to adulthood is a great way to figure out the precise identification.

2 **Prepare a home.** Caterpillars usually don't need much, living happily in any sort of box as long as there is food. A transparent container is best so you can easily see inside, and air holes are usually not necessary—insects respire very slowly, so opening the container to put in fresh leaves every couple of days will let in plenty of fresh air.

3 **Collect your caterpillar.** The easiest way to do this is to cut off the leaf or branch it is feeding on and place the whole thing in your container. Avoid touching the caterpillar as you can easily harm it and there are a few species that produce a painful sting.

4 **Keep a good supply of food.** Check on it every single day, and add fresh leaves from whatever plant you found it on. Some species of caterpillar can feed on multiple different types of plants, others feed on only one particular plant species. It is best to stick with feeding the caterpillar leaves of the plant you found it on, if possible.

5 **Provide a place to make a cocoon.** Some species hang from a twig, some roll themselves up in a leaf, and others like to burrow down under leaves or in the ground to pupate. A quick search online should tell you how your particular species pupates, but if you can't find any information—or don't know what species it is—give it a range of options. Put some soil and leaf litter on the bottom of the container, add some twigs and branches, and watch to see what it does.

6 **Let it transform.** Some caterpillars will stay in their cocoon just a few weeks; others will hibernate through the winter in their cocoon and emerge in the spring. If you can find information on the species online, give it the conditions it needs. If you can't find anything, or aren't sure exactly what species it is, place the cocoon outside where it can experience a natural range of temperatures and wait to see when it emerges.

7 **Release the adult.** When the caterpillar has completed its transformation, it will push out of the cocoon, spend a moment flexing its wings, and then fly off to find a mate and lay eggs for the next generation.

A chickadee has captured a caterpillar to feed its growing babies back at the nest. There are many useful apps and field guides to help you identify the different species found in your garden.

IDENTIFY EVERYTHING

The next step is to identify the specific birds and insects around you. This has never been easier thanks to modern smartphone technology. You can still go the old-school route of buying a stack of field guides specific to your region, but apps can make the process of identification a lot easier. There are many apps to identify plants and insects if you snap a photo of them, and even apps that identify birds just based on their songs. You may not be able to figure out exactly what every single insect you see is, but you'll quickly learn many new names and be able to identify a whole host of new neighbors every time you step outside.

KEEP A LIST

One fun way to build on your time observing in the garden is to keep a running list of everything you see. This is a practice taken directly from bird-watching, and there are many sorts of lists you can make. You could keep track of every different species you see in your entire life anywhere you go, or perhaps you could just focus on the species you find in your yard. You might even want to make a new list each year and see how many species you can track in a single season. However you choose to track them, and whether your notes are kept in a notebook, on your phone, or in a carefully organized spreadsheet on your computer, keeping a running list of the things you see will help you realize the diversity and richness of the world around you.

Every creature in your garden has a story to tell. Whether it's a monarch traveling for thousands of miles to lay eggs on its host plant or a beetle fighting to the death for a chance to mate, nature has some incredible stories to tell.

Choosing to incorporate a water garden into your landscape can bring a whole host of new creatures to your garden, including damselflies like this female ebony jewelwing.

LEARN THEIR STORIES

After identifying and listing everything you see, the next step is to take the time to learn more about the creatures on your list. So much of the beauty of nature is in the intricate lives that everything around us lead. Monarch butterflies, to take a classic example, are beautiful, but the real reason they capture so many people's hearts is because they make an epic cross-continental migration every fall and spring.

Once you start learning, the more you'll realize that every little fly, beetle, and butterfly has a similar story. They migrate incredible distances, hibernate through bitter cold, fight elaborate battles to win mates, and generally live lives with enough drama to fill a soap opera. It has never been easier to learn about the creatures you see in your garden. There are plenty of great reference books at your local library to help you learn, but a quick search online provides gobs of information right at your literal fingertips.

EXPAND YOUR VIEW

As you learn more about what's living in your garden, you can also start to learn about what *doesn't* live there now, but could if you create the right conditions. As you make more ecologically sensitive choices in your landscape, you'll see more and more species calling your garden home, but sometimes there are specific plants and conditions you'll need to provide to attract a particular species.

There are a lot of different ways to see what is missing from your landscape. In most regions, there are guided nature walks at local parks where experts will introduce you to a host of beautiful and fascinating local insects and birds. Jot down the ones you haven't seen at home, and with a little research, you should be able to find what they need to take up residence. You can also look at local field guides—paper or online—and see what species are present in your general area but, for whatever reason, you've never seen in your home landscape. This may mean planting a specific host plant for a butterfly or maybe making the right conditions for a specific species of firefly to lay its eggs.

Cats and the Subsidized Predator Problem

Cats are wonderful, adorable pets, but unfortunately, they can also wreak havoc on the ecosystem if you let them roam freely outside. Your cat may seem sweet and like not much of a hunter, but studies of tracked cats allowed to run free outside have found them to be extremely effective predators, killing huge numbers of birds, rodents, and even reptiles.

Predators are, of course, a natural part of healthy ecosystems. Foxes, coyotes, owls, and hawks also kill and eat many other animals, but they never have quite the impact of a pet cat because, unlike wild animals, cats are *subsidized predators*, whose survival does not depend on hunting success. If a fox or hawk eats too many of its prey species, it starts running out of food, causing the predator to have fewer babies or migrate to a new territory with a new food supply, allowing local populations of prey species to rebound. Pet cats, on the other hand, come home for a nice bowl of cat food, allowing them to keep on hunting your local population of songbirds just because they can. That means a pet cat can have a serious, long-term impact on wildlife populations in a way that wild predators never do.

All of this does not mean you can't enjoy having a wonderful pet cat in your life, of course. The simplest solution is to keep your pet cat indoors with plenty of toys and play time to keep it active and happy. Some cats can be trained to walk on a leash so they can still enjoy the outdoors while supervised. And, if you already have an outdoor cat, there are steps you can take to reduce their ability to hunt. A study published in the *Journal of Zoology* found the simple addition of a bell on a cat's collar can be 50 percent effective in protecting birds. There are other studies that suggest large, very colorful collars can also help birds spot cats and avoid being caught. It is also critical to keep cats indoors during late spring and early summer when many baby birds are first leaving the nest, and during the spring and fall migration seasons when many more birds may be passing through your landscape. Many cat lovers opt to install "catios" and other enclosed outdoor structures so their cat can spend time outdoors without risking harm to the cat or the birds it may hunt.

Even more damaging than pet cats are feral cat populations. If you have feral house cats in your community, the first step is to capture and sterilize the cats in the colony. If you provide food to a feral cat colony while some of the cats are still reproducing, you will simply create an ever-expanding subsidized predator population capable of devastating local bird and small mammal populations.

Taking these actions is the right thing to do for birds, but it's also the right thing to do if you love cats. Indoor cats live significantly longer than outdoor cats and protecting the birds from your cats will also mean protecting your cats from coyotes, cars, and other dangers of the outside world.

YOUR IMPACT *on the* WEB OF LIFE *in Your* GARDEN

WE ALL KNOW THAT the little choices we make every day have a big impact on the larger environment around us. Choosing to bicycle rather than drive to work on a summer day helps reduce smog and ground-level ozone pollution. Bringing reusable bags to the grocery store keeps a few more plastic bags from being made and used. Reducing the amount of meat and dairy you eat cuts way down on your contribution to climate change, habitat destruction, and pollution from animal waste. But, sometimes, it can feel futile, as you pack your reusable bags at the grocery store with low-impact, plant-based foods, and the person behind you is rolling up with a cart full of beef and cheese wrapped in plastic. Big problems like climate change require us all to do our part, but they also require government and industry leaders to make choices that we, as individuals, have little control over.

Your home garden is one place where you can individually make decisions with big impacts, even if everyone else around you chooses something different. Creating habitat for wolves and bison requires big efforts from all of us, but you, in your own backyard, can make choices today that help endangered species of insects that are equally as important as wolves and bison.

So, what actions should you take to ensure your garden is a vibrant habitat? There are many different pieces to the puzzle, but here a few basic principles of primary importance.

Small changes in how you care for your landscape and what plants you grow there can pay big dividends for a diversity of wildlife.

Woodland edges like this one filled with ferns and Indian pinks (*Spigelia marilandica*) host different plants and animals than dense woodlands or open grasslands. Create a diversity of growing sites that straddle different growing conditions if the site allows.

DIVERSITY

This is a theme we'll repeat many times in this book. Creating a diverse landscape is the single most important thing you can do to fill your garden with life. Many studies have looked at the number of species of insects found in a landscape, and pretty consistently, they find the most important contributing factor to a diversity of insects is the diversity of the landscape itself, which means you want a landscape that contains a diversity of shapes and forms of plants, a diversity of bloom times, and so on. If your landscape plants are all the same, the number of things that can live there is limited. Some insects like to live in grasses, others in bushes, and others in trees, so the more different types of plants you have, the more different types of insects you'll support. As an added bonus, diverse landscapes are fun to live in as well. A flat yard planted with lawn grasses supports only a handful of species and is also pretty boring. But mix that up with some variation by adding a small pond, some trees, shrubs, and a range of plants species, and you'll have something that is beautiful and wonderful to live in for you and a huge number of other animals.

Before we get into the specific sorts of diversity you should foster in your landscape, remember to think outside your property as well. Birds and insects don't respect property lines, so if you look at what is going on in properties around yours, you can also help create diversity at the neighborhood level. If your whole neighborhood is open and sunny, planting some big shade trees in your space will create a whole bunch of new habitat. You may not have room for every variation of plant life on your property, but focusing on being a little different than what is around you creates more diverse spaces for supporting life. Ecologists have found that "edge zones"—the places where different habitats touch or overlap—are the most consistently diverse and vibrant locations. Grasslands host one set of species, woodlands another, and the edge where they touch hosts both of those species' groups, plus a whole bunch of species that require the traits of both. The more edges you can create within your own landscape—and where your landscape connects with others around you—the better.

PLANT FORMS

There have been research studies comparing different home-scale landscapes to see which factors in their design and layout have the biggest impact on the diversity of insects and other life found in them. There are a lot of things that matter, but several of those studies found the most important factor to be the diversity of plant forms. This means a mixture of trees, shrubs, perennials, and grasses is best. Each form and shape of plant provides different places for different organisms to live, as some prefer open sunny areas and others the dense growth of shrubs or grasses.

Combining multiple plant forms in the same garden space is especially powerful because many birds and insects like to make use of different plant types. Many birds, for example, love to nest in the safe, concealed depths of a twiggy shrub, but venture out into open sunny areas to feed. The same is true of many insects as well. Often what is missing in traditional home landscapes is the shrubby middle layer. The typical home landscape has lots of open grassy areas and some tall shade trees, but nothing in between. Adding dense, shrubby intermediate-sized plants creates a more diverse habitat.

PLANT SPECIES

In addition to different shapes and sizes of plants, having a mix of different species of plants is the next biggest factor in building a diverse, life-filled habitat. The classic example of why this matters so much involves caterpillars and their host plants. Each species of butterfly and moth has a specific list of plant species it can lay eggs on and munch on in its caterpillar stage. Some species are generalists with a long list of possible host plants, and others can only live on a handful of species. And in many cases, these insects require at least two plant species to complete their full life cycle; one as a caterpillar and the other as an adult.

Oaks are powerhouse hosts for literally hundreds of different species of butterfly and moth caterpillars, but as fantastic as they are at providing food for the caterpillar stage, they have small, wind-pollinated flowers that produce no nectar at all to feed adult butterflies and moths. So, all the oaks in the world won't support very many species unless you also have flowers from other plants for the adults.

Plants come in many forms, and including a diversity of those forms in your garden is essential for supporting wildlife. Mix it up, and include trees, shrubs, perennials, grasses, and ground covers.

BLOOM TIMES

Flowers that provide nectar and pollen are a critical source of food for a huge range of insects. Ensuring you have flowers blooming over the longest possible stretch of time helps support the widest range of different pollinating insects. In cold climates, it is of course impossible to have flowers blooming through the winter, but pretty much any time the temperature is above freezing, there are flowering plants you can be growing.

One of the most common mistakes gardeners make here is focusing overly on spring-blooming plants. It is easy to do. Spring is when most of us get most excited about gardening, and when garden centers and nurseries conduct the overwhelming majority of their business. When we go to the nursery, we tend to grab everything in bloom right at that moment because it looks beautiful. Plants not yet in flower tend to be left on the nursery table. Plants like asters and ironweeds, which bloom gorgeously later in the season, just look like boring green blobs in the spring, making it easy to overlook them.

Throughout the summer and fall, walk through your garden and see what is flowering. If there is a lull in the bloom, take that as a cue to go to a nursery and look for something in flower at that very moment to add to your garden, or visit a local public garden where you can walk around and see what plants are flowering then to add to your to-buy list next spring.

Spring gardens are full of promise. Be sure to include some early spring blooming trees, shrubs, and perennials in your design.

Oak trees, when combined with other plant species growing around them, can provide food for hundreds of different birds, mammals, and insects in various life stages.

Summer really brings on the blooms, especially in sunny landscapes. Encourage a diversity of pollinators by planting a collection of plants with different flower shapes and colors.

Flower-filled fall gardens are of particular importance to insects who need to fuel up on pollen and nectar before overwintering or migrating. And don't forget to include plants with seed heads enjoyed by birds.

SUN AND SHADE

Having a mix of sun and shade in your garden creates diversity in two different ways. There is, of course, a long list of different species of insects and birds that prefer sunny or shady habitats, and another, possibly even longer list, of species that like to have access to both conditions. But, because different plants thrive in sunny and shady environments, having both in your garden is an easy way to bump up your species and plant-form diversities as well.

Shade trees are the first and most common way to create shaded areas in your garden, but they are not the only ones. Your house and other structures make shady areas on their north sides, and in hot, dry desert climates where large trees don't thrive, large rocks and boulders are often a critical source of shade for many plants and insects.

TOPOGRAPHY

Topography in a garden doesn't mean mountains and valleys. Though, of course, those are major drivers of a diversity of life in the world, they are just a bit beyond the scale of the average garden and gardener. But even small hillocks, berms, and low areas can create a wealth of interesting habitats.

If you live in a recently developed neighborhood, the natural humps and bumps and minor hilly variations of the landscape were probably completely removed by bulldozers during the construction process to create perfectly smooth, level, green lawns. Thankfully, putting back some of that normal variation is pretty easy. Mounding soil into low raised beds or berms, or digging out slightly lower areas can pay off in terms of creating diverse habitats. Low areas tend to stay wetter, which some plant and insect species love, while other species gravitate to higher, drier spaces. And some want to do both. A great example of an insect that uses both areas are ground beetles. These little guys like to rest in raised, slightly dried soils, especially in dense clumps of grass, but then they go to lower, wetter areas in search of their favorite food: slugs. So, if you have a generally wet garden where slugs are a problem, adding some raised, grassy areas for these beetles can increase the diversity of your garden habitat and reduce some of your slug damage.

Varied topography is important for creatures, but it can sometimes make a site challenging to garden. This gardener built terraces into a steep slope to prevent erosion and make the garden easier to tend. 

Even a small water feature brings big benefits to an ecological garden.

Ground covers act as a living mulch, shading the soil to reduce irrigation needs and out-competing weeds. This gardener has opted for an edible ground cover of strawberries.

WATER

Water from streams, ponds, lakes, and rivers is central to nearly every thriving ecosystem. Birds seek these water sources out to bathe, mammals seek them out to drink, bees land on rocks and pebbles to sip water safely, and butterflies suck moisture from the wet soil at the water's edge. If there is no natural waterway near your landscape, adding a small artificial one can amplify the diversity of living things visiting your garden space. This body of water doesn't have to be huge. You can, of course, dig a pond or install an elaborate water feature, but even something as simple as a small dish of water with pebbles in it so insects have a place to land while they drink can attract a shocking number of living things to your garden. This will have a huge impact year-round, but it is especially important during hot, dry weather when natural water sources are hard to find.

GROUND COVER

A final—and often overlooked—source of diversity in your garden involves how you cover the ground. For ease of maintenance, many gardeners keep their ground continually covered with a thick layer of mulch. As you'll know if you've ever dug through a mulch layer, there are many, many things that live within it. But there are also other insect species, such as ground-nesting native bees, that prefer open, sandy soils to create their brood chambers in. Providing a diversity of soil covers is a great idea for supporting both categories of critters. Some areas of the garden can be covered with a thick layer of plants, and others with mulch, but you should also consider letting some open, dry, sunny areas be just bare ground.

PESTICIDE USE

PERHAPS THE SIMPLEST way to increase the number of living things that visit your garden is to stop spraying the things that kill them. Lumped together, insecticides, herbicides, and fungicides are called *pesticides* because they kill all manner of pests, from insects and fungi to weeds and mites. But unfortunately, these products can't tell which insects we like and which we want to kill. The sad reality is that most pesticide treatments intended to kill a pest plant, insect, or fungus harm a whole host of other things that aren't causing any problems at all. And, of course, nothing is strictly a "pest." We call insects we don't like pests and plants we don't like weeds, but even extremely annoying cabbage worms serve as an important food source for baby birds, and many plants that previous generations of gardeners considered weeds are now being rediscovered as beautiful native plants well worth cultivating.

But the truth is that some insects are extremely damaging, and we do sometimes need to take action to control them. Mosquitoes and ticks are not just annoying: They can spread serious diseases. Tomato hornworms can decimate the tomatoes in your vegetable garden in no time, and sawfly larva can kill a young pine tree that was planted to create shade and increase the diversity of your landscape. The old way of gardening was what is sometimes called *spray and pray*—get some high-powered insecticide, cover the garden with it, and hope for the best. Today the preferred technique is IPM: integrated pest management. This technique acknowledges that there are pests we need to manage, but it involves managing them with a lot of information and planning and then taking the smallest, safest interventions possible to control the issues.

In general, native pests of native plants, such as these dogwood sawfly larvae on a shrubby dogwood, seldom lead to plant death as the two evolved together and damage is not typically extreme. Introduced pests tend to be more problematic in the landscape.

IDENTIFY THE PROBLEM

The first and most important key to solving a pest problem—and the first step in integrated pest management—is knowing the pest you are dealing with. Knowledge is power here. If you only know you have bugs eating your plants, then it is hard to make an intelligent plan to control that "bug" without harming all the other living things in your garden.

Getting a proper identification can sometimes be tricky, but start by taking advantage of the camera on your cell phone to snap a bunch of clear photos. Often just typing the name of the plant and a description of the insect (e.g., green worm on cabbage) into an online image search will allow you to find a picture that matches what you are seeing in your garden and put a name to the pest. Another fantastic resource is the Extension Service. Nearly every state in the United States has a helpline staffed by trained Master Gardener volunteers who can help you figure out what your pest is, especially if you can share a picture of the actual insect, not just the holes in the leaves it left behind.

When you have a name, read as much as you can about the pest. This will give you the information you need to take the next steps.

How much damage is too much damage depends on the gardener and the plant. Some pests, such as the introduced viburnum leaf beetle, can cause significant damage that affects long-term plant growth. Other pests cause damage that is largely aesthetic and therefore should be more tolerable.

DETERMINE YOUR DAMAGE THRESHOLD

Some level of damage in the garden is just fine—welcome, even. Plants are the basis of the food chain because other things eat them, so if you have a healthy garden ecosystem, you should expect—and even be happy—to see some holes in leaves. This means other creatures are thriving in your garden, too, which is exactly what an ecological gardener wants. The question to ask yourself is how much damage you can tolerate. If you have a caterpillar or two munching on a few leaves, that's okay. If, however, a plant is being mowed to the ground, that might be too much damage to tolerate. In that case, maybe it's worth choosing a different plant to grow that's more pest resistant.

If cabbage worms are eating all your kale, it might be unacceptable. But if you don't really like kale all that much, then maybe it's fine and you're willing to let the worms stay and become food for baby birds. Every gardener will have different ideas about what an acceptable level of damage is, and the damage threshold for a plant may change with time. Newly planted gardens with young plants may need more protection while they get established, while mature specimens can more easily bounce back from a little damage. Deciding how much damage is too much will help you decide when you need to take action to control a pest.

RELY ON NATURAL PREDATORS

One of the many benefits to creating a diverse garden habitat is that some of the many new species you'll be inviting will be predators. For every insect that is eating your plants, there is another insect that wants to eat *it*. Most pest problems in a diverse, healthy landscape will be self-limiting because the pests quickly become prey and never reach the level of a serious problem. There are exceptions, of course. The most serious problems are usually invasive pest species recently introduced from other parts of the world. In their native territory, every pest species has a host of other insects that evolved to prey on it, but when these damaging insects are moved to a new continent where there are no natural predators, they reproduce wildly with nothing to keep them in check. Luckily, with time, these new introductions tend to get under control as native insects and birds begin to recognize them as food. Government and university agencies also sometimes research and introduce predators from the pest's home continent to help manage them. When new pests arrive in your garden, you may need to take some action to keep them under control until the local ecosystem can adapt to this new visitor.

There are thousands of species of predatory insects that help control common garden pests, such as this spined soldier bug nymph feeding on young tent caterpillars.

The best way to employ natural predators in your home landscape is to create a healthy, diverse ecosystem to naturally support them. It is possible to purchase predatory insects and mites to add to your garden, but generally they are not a practical solution for most home pest problems (see sidebar on page 81). In addition, there are serious problems with some of the most commonly sold ones. Ladybugs are famous for eating aphids, but the species most commonly sold to gardeners (the convergent ladybug, *Hippodamia convergens*) is unethically harvested by collecting the adults from the spaces where they hibernate and packaging them into containers for shipment. There are many different species of ladybugs, and the best for controlling your pests are the ones that are already in your garden. So don't just purchase ladybugs because they are a "good" insect; rather, research your specific pest and see if there are biocontrol agents recommended by reputable sources like University Extension sites. Usually, the most effective ones to add are things like mites, nematodes, and fungi, not large insects that will easily simply fly out of your garden after you release them. It's also worth noting that any beneficial insects you introduce to your garden should be species native to your area. For example, praying mantid egg cases are often sold at garden centers for release into the garden, but these are either a Chinese species (*Tenodera sinensis*) or a European species (*Mantis religiosa*), not a North American one, and so should be avoided.

Your BFFs (Bug Friends Forever)

Among the insects in your garden, you'll find friends, foes, and those just passing through. Your friends, the beneficial insects, provide invaluable services in the garden ecosystem.

Pollinators often get the most recognition in the beneficial insect category—and for good reason. But the health of your ecological garden also relies on hundreds of other insect species that do heavy lifting in the form of pest control. You most likely already know about the big, obvious ones, like ladybugs and praying mantids, but some of these beneficial creatures are less charismatic. Some live in the soil and others are so small you probably don't even realize they exist. Their size doesn't dictate their appetite, however.

A 2006 study estimated that beneficial insects provide agricultural pest control services to the tune of $13.6 billion each year. Research is also being done on using beneficial insects to control invasive plant species, further emphasizing the essential role of bugs in our ecosystem. In agriscience speak, this is called *biological control*. If pepper farmers can rely on *Diglyphus* parasitic wasps to control leafminers in their crop fields, backyard ecological gardeners can trust in friends like this too.

Beneficial insects in the pest control category fall into two groups based on their way of operating:

- Predatory insects seek out and devour their prey. Predatory insects tend to be larger, and both their juvenile and adult stages are of great value.
- Parasitic insects parasitize their prey. Adults lay their eggs on or inside the prey's body, and when the larvae emerge, they eat the prey from the inside out. You may also hear these called *parasitoids*, as in parasitoid wasps.

A brown lacewing larva ready to capture and eat an aphid is an example of a predator.

This tiny wasp is about to lay eggs in nearby aphids: a classic example of a parasitoid.

ENCOURAGE THE BENEFICIALS

The work of beneficial insects may seem like something from a sci-fi blockbuster, but they're real and they're hungry for the pests in your garden. They are the film characters you want on your side, but how do you encourage them to take up residence in your garden?

One way beneficial insects find their way to your space is simply through the presence of their prey. Insects migrate to their food sources.

Another way they find your space is via the conditions you create. Like all living things, they need water and shelter, or habitat, for survival. Also, many of these critters aren't purely insectivores; they also need the carbohydrates in pollen and nectar. Include diverse sources of pollen and nectar in your garden as another attractant.

Alongside the work you do to attract the right insects is one obvious but sometimes overlooked deterrent. Don't use pesticides if you're trying to create a welcome environment for beneficial insects. Pesticides—even organic ones—can take out the good bugs along with the bad.

INTRODUCE THE BENEFICIALS

While rolling out the all-natural welcome mat for beneficial insects is the ideal, there are some insects that are slow to arrive, particularly the microscopic gang that lives underground. In this case, it's possible to purchase and introduce beneficial insects yourself.

You may debate whether introducing an insect to an environment is an ecological act. If the beneficial insects are native to your region and your garden is set up to accommodate them with an already-present population of prey and supporting food sources and habitat, you could reason that you won't do harm with this introduction. On the other hand, by artificially introducing beneficial insects to your garden ecosystem, you're preventing the native population of these natural-community members from moving in on their own. If they show up and their preferred prey is already under control, they won't stay.

Also consider the management involved in making this introduction successful: The pest population timing and weather conditions need to be just right (not to mention you have to be present to receive the package of live insects so you can immediately deploy or properly store them). Plus, all of this costs money. Biocontrols are an in-demand business. Looking specifically at the predatory beneficial nematodes often used to manage soil-dwelling pests like grubs and borer larvae, there are a few additional considerations. You'll need enough bare soil to apply them, as they won't burrow through mulch. After applying them you'll need to water them, usually using a watering can, at regular intervals until they become established in the ground.

Larval ladybugs might look scary, but they are your BFFs in the garden. This one is making lunch of an aphid on a leaf.

KNOW YOUR FRIENDS

It's nice to recognize there's a whole community of insects coming and going from your garden, making the ecosystem as vibrant as it is. In the case of beneficial bugs versus pests, it's important to know who's on your side. Some of these beneficials are, truly, weird looking. When you spy your first ladybug larva, first impressions may lead you to believe it is not a good bug. It looks like a tiny black-and-red armored vehicle (the perfect disguise for attacking all those aphids!). And to be able to recognize beneficial insects at all their life stages, from eggs to larva or nymph to adult, is equally as important.

Get your hands on a reputable guide to garden insects, ideally one that's specific to your region. This might even be a resource available from your Extension Service, in print or online.

The small computer you carry in your pocket is a whiz at making identifications as well. Several entomological and photo-recognition apps can zero in on insect ID in a split second. And remember, you can also do a simple online search for the bug's characteristics and pull up all you need to know on a plethora of websites with photos to confirm the identification. Cross-check multiple sources when you're just starting out or when you encounter a new-to-you insect.

Meet a few of the BFFs that could be in your area, learn their favorite foods, and read about how to lure them to your garden.

BENEFICIAL INSECT	PREY	MODE OF ATTACK	ATTRACTANTS
Assassin bugs	Leaf-footed bugs, caterpillars, leaf beetle and sawfly larvae, and other insects' adult and nymph stages; some species may feed on beneficial insects	Both adults and nymphs stalk their prey, inject venom using their tubular mouthparts, and suck out the liquified contents.	Plant goldenrod, dotted horsemint, buckwheat, and more. Provide dense foliage for their hunting grounds.
Damsel bug	Aphids, beetles, caterpillars, mites, thrips, lygus bugs, and others	Adults and nymphs use needlelike mouthparts to suck prey's body fluids.	Plant caraway, cosmos, fennel, alfalfa, marigolds, and more.
Ground beetles	Wireworms, maggots, slugs, snails, ants, aphids, caterpillars, armyworms, cutworms, cucumber beetle larvae, Colorado potato beetle larvae, and more; may also eat weed seeds	Larvae and adults can eat their weight in prey daily, using sharp mouthparts to chew up prey.	Provide rocks and mulch as habitat, along with planting native clump-forming grasses. Turn off outside lights at night, as they'll distract beetles from the garden.
Ladybugs (a.k.a. ladybirds and lady beetles)*	Aphids, mealybugs, scale insects, whiteflies, and other soft-bodied insects	Adults lay 20 to 1,000+ eggs each season on leaves and stems where prey are present. Adults and larvae use chewing mouthparts to eat prey. One ladybug can eat 5,000 aphids in its lifetime.	Give them space to overwinter in fallen leaves and dead trees. Plant shallow flowers with staggered bloom times, including sunflowers, yarrow, dill, and sweet alyssum.
Steinernema glaseri nematode	Japanese beetle	This nematode parasitizes beetle grubs underground.	Introduce nematodes from an outside source into your soil.
Syrphid flies (a.k.a. hoverflies and flower flies)	Aphids, mealybugs, thrips, and other sap-sucking insects	Adults lay eggs on leaves near prey. The eggs hatch within 3 days. Over 1 to 3 weeks, each larva can eat up to 400 aphids each.	Provide pollen and nectar-producing plants. Leave fallen leaves and plant hedgerows for overwintering.
Tachinid flies	Corn earworms, cabbage worms, cabbage loopers, cutworms, armyworms, stink bugs, squash bug nymphs, beetle and fly larvae, beetles, and others	Adults deposit eggs on foliage or on the prey itself. Larvae feed from within the prey's body.	Plant California buckwheat, pennyroyal, parsley, phacelia, tansy, and more.

*Of the hundreds of types of ladybugs in North America, a small handful, like Mexican bean beetles, are pest insects rather than beneficials. Further complicating things, each life cycle stage looks completely different than the one before it. Get to know the species in your area so you can distinguish the good from the bad.

Handpicking is a simple, targeted way to eliminate a pest, in this case a delphinium budworm.

CHOOSE THE SAFEST CONTROL

Even when you give beneficial insects time to arrive and work their magic, you will occasionally have a pest outbreak they can't manage. When this happens and you need to take action, always choose the safest, least damaging intervention possible. In general, that means choosing a control option that eliminates the pest but nothing else. The spray bottle of insecticide at the hardware store that promises to kill a long list of insects and pests is the worst option. Any spray that is labeled to manage ants *and* aphids *and* ticks *and* fleas *and* spider mites *and* mealy bugs will also wipe out honey bees and beneficial beetles and butterflies. Remember, this is true whether the insecticide is synthetic or organic.

Synthetic pesticides (often also called *conventional pesticides*) are created from man-made chemicals formulated in a laboratory, while organic pesticides are based on naturally derived materials, such as plant derivatives, minerals, and oils. Many organic insecticides are more targeted and kill fewer insects, and when they are used as part of an integrated pest management program, they greatly reduce and can even eliminate the need for synthetic pesticides. There are, however, also broad-spectrum organic insecticides that kill a whole swath of organisms that are not causing any problems at all. There are many good reasons to choose organic options, but don't simply rely on the word *organic* on a package and assume the product will not harm organisms beyond the one you are targeting; take the time to read the whole label thoroughly and choose the most targeted, least toxic option you can. Many ecological gardeners opt to use no pesticides at all, instead relying on prevention and manual pest removal.

To that end, sometimes the most targeted option is also the simplest one: your fingers. In a small space, simply picking the pests off with your finger and crushing them or dropping them in a bucket of soapy water is a great first line of defense. It can be tedious—and gross—but it is easy to target just that pest and nothing else. Manual removal is a simple way to knock back an outbreak of pests without harming the other living things in your garden.

Another less-toxic option to look for are bioinsecticides. These products are made from living things, like nematodes, fungi, or bacteria, and they usually have extremely narrow ranges and attack the targeted pest and nothing else. Milky spore, a fungal disease that attacks Japanese beetle larva, is a good example of this, as are nematode sprays that can control vine weevils in the soil.

In addition to the type of control you use, the way you apply it can make a big difference. Bt toxins, for example, are an organic insecticide produced by the *Bacillus thuringiensis* bacteria. There are different strains that target different types of insects—one strain targets only flies, for example, while another kills caterpillars. The fly-killing strain is very effective at eliminating mosquito larvae. If you spray it over your whole garden, you'll kill a lot of other flies, most of which are completely harmless and many of which are great predators of other pests or fantastic pollinators. But if this product is added only to standing water, as is commonly recommended, the insecticide can be delivered only to the water-dwelling mosquito larvae, leaving all the other flies unharmed.

A similar approach can be used to control ticks by spraying insecticides on cotton balls and placing them in tubes out in the garden. Mice, voles, and other rodents collect the treated cotton to line their nests. There, the insecticide kills ticks on the mice and their babies, very effectively knocking back the tick population while leaving other insects unharmed. In this case, the fact that the insecticide used is pretty powerful and broad spectrum matters less because it is being delivered in a careful, targeted way.

These targeted approaches to controlling pests are not only better for the larger ecosystem, they are more effective as well. Carefully delivering an insecticide into rodent nests to knock out ticks will reduce their populations far more than just buying a bottle of "It kills everything" at the hardware store and spraying the whole garden with it. It will be cheaper as well. All it takes is a commitment on your part to spend a little time learning about your specific pests and how to treat them in a safe, effective, and targeted way.

PROVIDE FOOD

ON A PRETTY FUNDAMENTAL LEVEL, if there is nothing to eat in your garden, nothing can live there. Insects and other living things in your garden get their food from a variety of sources, be it nectar and pollen, plant leaves, decomposing plant material, or other insects. Making sure all of these animals have a steady supply of food is key to building a healthy ecosystem in your landscape. Let's talk about how to provide food for some of the most popular residents in an ecological garden.

POLLINATORS

These are the easiest garden residents to love. Pollinators visit flowers and feed on nectar or pollen. As they move from flower to flower to feed, they spread pollen between the blooms, fertilizing them so the plant can produce seeds. Or, to put it another way, the plants bribe the pollinators with food in exchange for sex.

Provide plenty of nectar and pollen from a diversity of flowering plants to keep beneficial insects like this syrphid fly happy and healthy.

Plant-pollinator relationships are one of those pleasing parts of the natural world where both parties involved benefit, at least most of the time. There is quite a bit of cheating that goes on between plants and pollinators. Some flowers trick pollinators into visiting them but provide no food reward. Native lady slipper orchids, which are beloved for their showy, unusual blooms, are one of these plants. They lure insects in with their colorful petals and then use their oddly shaped blooms to push the bees through a specific route that results in pollen sticking to their head but nothing to eat. The cheating goes the other way too. Some bees do what is called *nectar robbing*—instead of going in the top of a flower to get nectar and getting dusted with pollen as they do, they land on the side of a bloom and chew through the petals to get at the nectar without having to go through the normal pollen-dusting route.

The most well-known pollinator is, of course, the honey bee, but they are far from the only, or even the most important, pollinator. Honey bees are not native to North America. They are very important commercial livestock, used to pollinate fields of vegetables and fruits, but in natural North American ecosystems they are not necessary and are even considered an invasive species. Feral and domesticated honey bees can sometimes compete with—and spread diseases to—our native bees, negatively impacting their populations.

Most of our 3,500+ species of native North American bees don't form large colonies that live for many years like honey bees do, making them easy to overlook. Some, like bumblebees, form small colonies that last for just one summer, and most are solitary nesting species, simply collecting nectar and pollen to feed a few babies that will carry on to the next year. These native bees, however, are just as effective pollinators as honey bees and—on the plus side—are much less likely to sting or be aggressive.

Different floral structures attract different types of bees. This false indigo (*Baptisia australis*) has lobed flowers that require a heavy pollinator to open them. Here, one is hosting both a bumblebee and a wool carder bee.

Hummingbird clearwing moths are a common summer pollinator on bee balm and other plants across much of North America.

Our many native bees are important pollinators, of course, but so is nearly every other group of insects. Wasps, beetles, butterflies, and moths are all critically important pollinators, and one of the most significant groups of pollinators is flies. Flies may be best known for buzzing around annoyingly and sometimes biting you, but only a few species of flies actually do that. There are over 16,000 species of flies native to North America. Don't let the few pesky species tarnish the reputation of this enormous group of insects; many of them are very important pollinators and others are pest predators.

All of these different pollinators have different preferences when it comes to the flowers they visit. Bees tend to prefer small purple and blue flowers; moths generally go for large, white, fragrant blooms; hummingbirds typically visit red, trumpet-shaped blooms; and flies and wasps prefer plants that produce big clusters of small flowers, like goldenrod. Ensuring that you have a diversity of colors, shapes, and sizes of blooms in your garden is one simple way to make sure you provide food for the widest range of pollinators possible.

Include native plants in your garden that attract and support a diversity of pollinators, from butterflies and bees to wasps, beetles, and flies.

Beyond having a range of flower types, you also need to make sure pollinators have other essential food sources as well, because, except for some species of bees, most pollinators don't rely exclusively on flowers for food throughout their entire lifecycle. Nectar is essentially sugary water, so though it is a good quick source of energy, the vast majority of pollinators require additional food sources—like protein—that come from elsewhere. It might surprise you to learn that some of the best pollinators are also predatory insects. They get their protein by hunting down pesky aphids and other problem insects and then stop by flowers for a quick sweet nectar snack. Many types of pollinators also require protein, whether from pollen or prey, to feed their developing young. Even hummingbirds, one of the most iconic pollinators out there, evolved their rapid, acrobatic flying abilities in part so they can catch insects in the air. These little birds do like to visit flowers for a quick hit of sugary calories, but they are primarily insect eaters, particularly when they are raising young that need a lot of protein to develop.

We now circle back to the main theme of this chapter: diversity. A healthy group of pollinators in your landscape will consist of many different species, not just a hive of honey bees. In order for this to happen, you need to provide these creatures with a diversity of food sources, both in the form of flowers and other insects to catch and eat.

HERBIVORES

If pollinators are the easiest garden residents to love, the herbivores are the hardest. These are all the critters that eat your plants, from aphids and caterpillars to rabbits and deer. These hungry beasts are the ones you are most likely to classify as pests, and they can be devastating to a beautiful garden. But it is also important to accept them as part of a healthily functioning ecosystem. When plants are eaten by insects and other herbivores, plant energy moves up the food chain to fuel all the beautiful diversity of life around us.

It doesn't, however, mean you have to accept a tattered garden eaten to the nub. In the vast majority of cases, herbivorous insects will nibble a little here and there, but not do serious damage, and if their numbers get too out of control, hungry predatory insects will come along and gobble them up. When the hungry herbivores do get to be too much, refer to the section on page 82 on controlling them safely. However, most of the time you won't even notice a few holes in your plant leaves unless you look super closely, and those few holes are a good sign that your garden is feeding the local ecosystem.

Most herbivorous insects are prey for someone else further up the food chain. Here, a thread-waisted predatory wasp has captured a pest caterpillar.

Small herbivorous mammals, like this rabbit, can damage garden plants. Proper fencing is key to humane control.

Harder to deal with are the larger herbivores like rabbits and especially deer. White-tailed deer in much of North America are an enormous problem because we have largely eliminated their natural predators—wolves, bears, and big cats like cougars, pumas, and bobcats—from the ecosystem, so deer populations have skyrocketed far beyond their historically normal levels, often with devastating consequences. There is evidence that deer overpopulation is remaking the whole structure of woodlands, clearing out species they like to eat and wiping out certain wildflower populations, and replacing them with species the deer don't eat. In the face of that, excluding deer from your garden will not only save you frustration, but provide a safe haven for deer-favored wildflowers like trillium and asters. Repellent sprays can work short term, but long term, the only effective way to exclude deer is a tall (6 to 8 foot [2 to 2.5 m]) fence.

Smaller herbivores like rabbits and groundhogs can be extremely annoying, but their populations are much better controlled by the expanding ranges of coyotes and foxes, as well as raptor populations, which are rebounding across North America since the banning of DDT. Generally, a high rabbit population means that, in a year or two, hungry predators will move in, and the rabbit numbers will crash. Permanent fencing can be critical to protect vegetable gardens from rabbits, and you might consider using temporary chicken wire fences around newly planted garden areas to let the plants establish before rabbits start browsing. However, most mature perennial and shrub gardens can handle a little rabbit browsing.

DETRITIVORES

These are the living things that are the easiest to overlook because they mostly live out of sight, feeding on leaves, branches, stems, and roots of plants after they die and start breaking down. Earthworms are perhaps the most famous of these animals, but they are just the tip of the iceberg. Countless insects, arthropods, worms, and other organisms live on the bits and pieces of plants at the end of their lifespans, while others feed on the fungi and bacteria that consume decomposing plant material. And, of course, all these living things then work their way up the food chain as bigger insects, birds, and mammals eat them.

Detritivores may not be as glamorous as butterflies and bees, but they are a huge way the plants in your landscape contribute to the local food chain. Best of all, they do it without harming the plants. Unlike herbivores, there is no balancing act required here to keep populations low enough to avoid significant damage. All you need to do to keep these creatures fed is to let as much plant material as possible decompose in place, rather than clearing it all away.

This is one of the places where doing what is right for your local ecosystem means doing less work—letting autumn leaves be instead of raking them up, for example. There are times, however, when leaving fall leaves or brown stems may not be practical, and we'll discuss these instances more in chapter 6. What's important to know now is that autumn leaves are an important food source for detritivores, so finding a way to keep them on-site is important. Unseen corners behind shrubs are great places to move this dead plant matter. There it will provide just as much food for detritivores as it did when out in more public view.

One of the most important sources of food for the detritivores in your landscape are trees when they reach the end of their life. A tree trunk is a huge supply of energy stored up over decades of photosynthesis, and there are countless organisms whose life cycles are built around the various stages of decay in trees. However, trees are also one of the trickiest things to let decay in place, especially on small properties. If you have a large property, and there are trees well away from where they might fall and hurt a person or damage buildings, the best course of action is to just let them stand. Standing dead trees—sometimes called *snags*—are home and food for many birds, insects, and fungi, many of which can live and feed nowhere else.

However, if a tree hangs over your house or a busy walkway, it simply isn't safe to let it decay in place. Just as with cleaning fall leaves, when at all possible, cutting a tree down and moving it somewhere safe to break down is your best option here. Fallen logs can quickly look beautiful in a woodland garden as they accumulate a layer of moss and eventually ferns and other bigger plants as the wood slowly decomposes. Even if you have to get a wood chipper and turn much of the tree into mulch, it will still enrich your soil and provide food for a lot of the life there.

Detritivores, like this pill bug, are essential ecosystem partners for breaking down decaying materials.

Let dead trees (snags) stand in the landscape wherever it's safe to do so. They provide a host of benefits to wildlife, including this hairy woodpecker.

PREDATORS, INSECT AND OTHERWISE

All of the insects and other animals eating leaves, nectar, and decomposing plants are themselves food for the next stage of the food chain. Everything from predatory insects to moles and songbirds gobble up worms and caterpillars and flies, and then, often, are themselves eaten by still larger predators. It is important to remember that any synthetic chemical pesticide used to control pests will be evermore concentrated in the bodies of predators as it moves up the food chain. The infamous pesticide DDT is a classic example of this. It had little impact in the bodies of songbirds and small mammals, but reached high concentrations in eagles, hawks, and other raptors, nearly wiping out whole species of these beautiful predatory birds.

In your home landscape, it may be tempting to put out poison bait to eliminate hungry voles or mice that are eating your plants, but that poison will move up the food chain and harm foxes, hawks, owls, and even sometimes pet cats and dogs. Respecting and protecting that often familiar, glamorous species at the top of the food chain means respecting and not poisoning the smaller species it depends on.

This sharp-shinned hawk stands guard over the garden as it hunts for mice and other prey. Keep raptors safe by not using poison baits for mice or voles. 

Why Not Just Put Out Food?

As we talk about how a complex, healthy ecosystem provides food for the species we love, you may be asking why you can't just feed the animals you like the most. There are a few reasons why this may not be such a great idea or may not work exactly as you might imagine. Bird feeders with seeds for songbirds and sugar water for hummingbirds are great ways to attract birds into where they can be easily observed, but these feeders are not a replacement for a healthy ecosystem. Virtually no species of bird feeds just on seeds—most need a steady supply of high-protein caterpillars and other insects, especially when they are laying eggs or feeding babies. Concentrated feeding stations can also bring birds into much closer contact than they normally would, sometimes encouraging the spread of disease and leading to conflict (not to mention concentrating them into a specific area, which makes them more vulnerable to predators like cats, see sidebar on page 66).

That isn't to say you can't put out bird feeders—they can be a wonderful way to learn more about the birds that call your garden home—but they are not a replacement for a healthy ecosystem full of diverse foods for birds to eat.

PROVIDING SHELTER

Including many different plant forms in your garden provides the greatest diversity of habitats for wildlife. Many plants play multiple roles: They provide nesting habitat; shelter from predators; overwintering sites; food in the form of pollen, nectar, fruits, or leaves; and so much more.

THE LIFE IN YOUR GARDEN not only needs food to eat, but it also needs a place to live. Every little nook and cranny of the natural world is a suitable home for a different sort of living thing. In general, approach providing good housing for living things by focusing on the same principle we've discussed before: diversity, diversity, diversity. If you have a mixture of many different plants with varying forms and shapes, you'll wind up also providing homes for a range of different living things. But there are also some specific types of habitats you'll want to make sure you include in your landscape.

DENSE VEGETATION

Tight, dense areas in shrubs and thickly growing perennials are prime habitat for many animals, especially small birds who feel safe from predators in such habitats. Unfortunately, these dense, safe habitats can also be prime locations for species we want less often in our gardens, especially rodents and herbivores like rabbits. That doesn't mean you can't include these areas in your landscape. It just means to be mindful of what may occur. There are many examples of beautiful, densely planted gardens that don't house rodents situated right next to buildings, but if you're worried about a potential problem, it may be best to strategically place these areas well away from your home and vegetable gardens to discourage potential rodent residents from wandering over to where you don't want them.

HOLLOW STEMS

Many perennials, and even a few shrubs, have stems that are either hollow or have a soft, pithy center, making them a perfect home for a range of different insects, including many species of native solitary bees. The specific insects that will take up residence in hollow or pithy stems depends on the size of the opening (⅛-inch [3 mm] of pith or hollowness is ideal) and your specific garden location, but leaving hollow stems standing, rather than cutting them down to the ground, creates habitats for a whole host of insects.

Hollow plant stems are used as brood chambers and overwintering sites for many species of native insects. Here, a smooth hydrangea (*Hydrangea arborescens*) stem is home for a small carpenter bee.

Some native bee species set up shop in holes in dead wood. Here, an orchard mason bee is making her brood chamber in a log.

Fall leaf litter is a common wintering site for many different insects, including congregations of the convergent ladybug.

LEAF LITTER

The layer of slowly decomposing dead leaves and stems at the soil's surface is prime living quarters for many insects. Lots of species make the leaf litter their full-time home, and even more retreat under a layer of dead leaves as a safe spot to hibernate through the winter. As we discussed in the section on detritivores, leaving dead leaves may not be practical everywhere in the garden, but making sure you have at least some places on your property where you can let leaf litter build up ensures that many insects have a safe place to hide, feed, and possibly hibernate.

When possible, try to avoid disturbing dead leaf piles once they have formed. Letting leaves sit in place over the winter and then cleaning them up in the spring may unfortunately lure insects to lay eggs or form pupae that are then harmed when the leaves are collected during a spring cleanup.

DEAD WOOD

Not only does dead wood provide food for detritivores, but dead trees, twigs, and branches also provide safe shelter for many things. From woodpeckers that dig out homes in dead snags to a whole host of beetles and ants, many living things find homes when you leave dead wood in the landscape. Practically speaking, however, letting every branch fall is not realistic for most managed landscapes. Dead branches can create trip hazards and look untidy and unattractive in the garden. If you have an out-of-the-way corner of your garden where you can relocate branches to decompose, great, but you can also get creative with ways to keep dead wood around while making its presence look more intentional.

Leafcutter bees are one of many solitary-nesting native bee species that build their nest cavities in bare soil. Leave sunny, exposed patches of ground unmulched and welcome them to nest.

One way is to create a *deadhedge*. Simply pound two lines of branches or metal rebar into the ground and place a collection of fallen branches in between them, pushing the branches down to make a straight line. Over time, your deadhedge will be built up to create a sort of informal wooden wall that can look quite attractive, and the branches will slowly decompose to provide both food and housing for a range of creatures.

BARE SOIL

While many insects prefer to shelter under leaf litter, as you now know, there are other insects that like to shelter or nest in the bare ground, so keeping some areas clear allows for a different set of garden residents. This reality can be a great connection with the more practical, aesthetic goals you have for your garden. Perhaps you can let leaves accumulate and dead wood fall in a more wild, back corner of the garden, and a more formal section in the front yard might be a good spot to keep meticulously clear of mulch to allow ground-nesting bees to make their homes.

SOIL LIFE

Decomposers, like this American giant millipede, help break down leaf litter and other plant debris to feed the soil and fuel future plant growth.

THE LEAST VISIBLE—and arguably the most important—living things in your landscape are all the organisms that call your soil home. An astonishingly diverse group of species live in your soil, most of which are overlooked because they are underground, where they are quite hard to see, and often microscopic. But just because they're hard to see doesn't mean there aren't a lot of them or that they aren't critically important. Just a teaspoon of soil is estimated to contain tens of thousands of different species of single-celled bacteria, nematodes, fungi, and other organisms. They are all busy living their complex lives, but they provide a lot of benefits to you and your landscape. Let's explore a few of their roles.

DECOMPOSITION

One of the biggest things the soil life is doing is breaking down organic matter in the soil. Without this process, we'd be wading through an ever-deepening layer of dead leaves, stems, branches, dead animals, and other wastes. Even more critically, plants wouldn't be able to recycle the nutrients of previous generations to fuel their growth. When you stop to think about it, it's a pretty magical process. Simply letting a leaf drop onto the ground means it can be broken down into its constituent parts, which are then taken up by plant roots and transformed into a whole new leaf. If only recycling plastic were so easy and efficient!

SOIL STRUCTURE

Soil structure is an easy-to-overlook concept, but a powerful one for creating soil that is a healthy place for your plants to thrive. Soil is primarily made up of tiny mineral components that are classified based on their size—sand particles are the largest, silt is midsize, and the smallest particles are clay.

The small particles of clay and, to a lesser degree, silt, can pack together so tightly that there is no space between them to allow air, roots, and water to move through the soil. Sandy soils, on the other hand, can be so open and loose that, though roots grow through them easily, any water and nutrients quickly drain away, creating very dry, infertile soil conditions.

Soil structure is when those individual soil particles are bound together to create little clumps called *soil aggregates*. If you've ever dug up soil and had it crumble apart into little granola-like chunks, you've seen soil with a good structure. The aggregates in soil with a good structure fit together loosely, allowing air, water, and roots to easily penetrate, but each individual aggregate acts like a little sponge, holding more water and nutrients than a loose, sandy soil. Basically, whatever soil you have, improving the soil structure is what will make it less likely to get waterlogged in rainy weather while enabling it to hold more water in dry conditions.

Good soil structure also makes it easier for plant roots to grow and access the water and nutrients they need. What creates good soil structure? You guessed it! All the tiny living things in your soil. Soil bacteria, fungi, and arthropods, along with organic matter and the action of plant roots, bind those soil particles together, creating coveted good soil structure.

SYMBIOTIC RELATIONSHIPS

Many of the living things in the soil have tight, mutually beneficial relationships with plant roots. The biggest group of these are the fungi called *mycorrhizae*. These fungi form networks through the soil and attach tightly to plant roots, where they act almost like an extension of the plant's root system. The fungal threads supply water to the plants, just like roots, and most critically, they're better than plants at extracting certain nutrients from the soil, so they almost act as living fertilizer. The plants, in return, share some of the sugars they create through photosynthesis with these fungi, giving them the food they need to keep growing and sharing.

In healthy soils, these mycorrhizae in the soil connect between the roots of many different plant species, and they can make an entire garden grow more vigorously as they collect water and nutrients from the soil and deliver them to plants. This makes the garden significantly more drought resistant, as each plant can essentially plug into the shared fungal network to help source water and nutrients.

FOOD CHAIN

All these living things in the soil are, of course, food for other living things. Microscopic organisms are eaten by springtails and nematodes barely bigger than they are. These creatures in turn feed larger insects, and so on up to the birds and mammals. Rich, vibrant soil life is one of the key foundations of a diverse, healthy ecosystem.

Small clumps of soil particles, called *aggregates*, determine how water and air move through the soil. These aggregates are held together by organic matter and soil organisms.

Fungal networks throughout the soil act as extensions of a plant's root system, mining nutrients and water for the plant in exchange for carbohydrates from the plant roots.

CARING *for the* SOIL LIFE

WE'RE SURE YOU CAN now see that soil life matters. A lot. But how do you care for it? There are a few key things to keep in mind to keep these organisms thriving.

MINIMIZE DISTURBANCE

Soil, in nature, mostly stays in one place. There are occasional disruptions from erosion, a tree blowing over, or an animal digging a hole, but those are exceptions. Humans, on the other hand, tend to be incredibly disruptive. In new construction, we bring in huge machines to dig and churn the soil; through poor water management we create damaging erosion; and some gardeners use regular tilling to control weeds, churning the soil over annually. Some of this disturbance is inevitable, but you should try to minimize it, and ideally, make it an infrequent event.

Getting a garden tiller out every year to turn over the soil can make it quite difficult for a healthy network of soil life to become established, as well as physically breaking down the oh-so-valuable soil structure. Using a tiller as a one-time event (in the process of transforming a lawn, for example, into a new garden bed) will damage soil life, but the network will quickly recover if it is left undisturbed after that.

AVOID FUNGICIDES

Soil fungi, particularly mycorrhizae, are so critical to soil health that you'll want to be particularly careful when using fungicides in the garden. Many of the fungicides that are used to control plant diseases can also damage soil fungi. In general, avoid using fungicidal sprays when you can, and try to choose disease-resistant plant varieties

No-till vegetable gardens skip annual tilling in favor of yearly additions of organic matter and layers of mulch.

Rather than hauling plant debris off to a compost pile, let it decompose in place. This broken-down plant material provides organic material to the soil as it decomposes. Its presence also helps reduce weeds and slowly builds good soil structure.

instead. But not all fungicides are created equal in this regard. Fungicidal sprays you apply to the leaves of plants generally have minimal impact on the soil, as long as you are careful not to overspray and let it soak down into the soil. Systemic fungicides, which are applied to the soil and then taken up by plant roots, have the biggest impact on mycorrhizae, and should be avoided. Caution should also be used with copper-based fungicides. These fungicides are classified as permissible for use in organic gardening, but because the active ingredient is the element copper, these fungicides never break down, and if used regularly, copper can build up in the soil. This buildup has been shown to harm the growth of mycorrhizae. Using a copper-based fungicide once or twice is unlikely to have any real impact on the soil, but relying on them on a regular basis is not recommended.

PROVIDE ORGANIC MATTER

The ultimate source of food for most of the living things in the soil is organic matter—any sort of old plant material that is decomposing. Simply letting leaves, stems, and flowers drop naturally to the ground and slowly break down is the easiest and most natural way to maintain healthy levels of soil organic matter (nutrient cycle section, pages 181 to 206). If your soil is starting with a severe deficit of organic matter, you may want to jump-start that process by applying a layer of compost over the soil. This will most often be necessary if your soil has been severely damaged recently, as is most often the case around new construction. When homes are built, the builders often remove and sell the topsoil, and replace it with the nutrient-poor sub soil they dug out while putting in foundations, which mean you'll be starting with almost no organic matter in your soil at all. While this is not as problematic for many native plants that prefer "leaner" soils, traditional ornamental garden plants may find these conditions challenging for establishment.

ADDING SOIL LIFE

One thing you *don't* need to do to create healthy soil is purchase commercial mycorrhizae or other soil organisms. These organisms are naturally present in soil and air, and if you provide the conditions for them to thrive by adding organic matter, minimizing tillage, and avoiding fungicides, they'll thrive and grow. Research has shown that adding commercial mycorrhizae products is only really useful when you're growing in sterile soilless media, like the potting soil used to fill containers. These products arrive in the bag nearly sterile, so adding organisms to the soil can be helpful. But in the ground, with natural soil, the beneficial soil organisms are already there; you just need to stop harming them and provide them with the organic matter they need to thrive.

Each plant evolved to grow in a particular soil type, so learning about your soil and choosing plants that naturally thrive in it is a big step toward success. A prairie-fire (*Castilleja* spp.) in a California seaside planting.

LOVE THE SOIL YOU HAVE

When gardeners start thinking about their soil, it is easy to get caught up in wishing you had "better" soil—richer, better drained, easier to dig, all of that. Though there are things you can do to alter your soil, in general the most sustainable option is to learn to love the soil you have and choose to plant the native plant species that will thrive in it.

Increasing organic matter by adding compost and letting plant material break down in place will improve the soil you have, but big interventions like hauling in tons of soil amendments or double digging the soil are expensive, labor intensive, and are unlikely to have the lasting impact you are hoping for. Whatever your soil is naturally, there are a whole host of plants adapted to thrive in those conditions. If you have very dry, sandy soil, there are many beautiful plants that demand those conditions and will be prone to rotting out if you manage to transform your soil to be richer and hold more water. And heavy, wet clay soils have their own plants that love those conditions.

If necessary, you can focus your soil-adjusting efforts to target a small, specific area. Dry, sandy soils are great places to show off a range of beautiful native plants, but they aren't the best for most common vegetable species, so perhaps you want to build some raised beds to fill with richer soil. Or maybe you love hardy cactus but have heavy, wet clay soil, so build a small raised sandy area to enjoy them. The more you can keep these efforts small and contained—and lean into your natural soil conditions everywhere else—the less work, expense, and even carbon footprint your garden will require.

This chapter is packed with information because of the many factors involved in ensuring that your landscape is a healthy, vibrant ecosystem. But it all comes down to a few basic principles.

The first is to learn about and celebrate all the different organisms—from fungi and insects to birds and mammals—that share your landscape with you. Next, focus on increasing diversity. The more different plants, different bloom times, different conditions, and different topographies your garden has, the more varied the insects and other animals that find their way to your garden will be.

Finding the right plants for your soil type is as easy as looking to plants that are indigenous to wherever you live.

And finally, any time you feel the need to intervene in your landscape, be it to control a pest, clean up a mess, or kill a weed, do it carefully and thoughtfully. Be a force for good in the garden's ecosystem, rather than a danger to the living things around you.

In the next chapter, we'll take a look at one of the most essential ways to enhance your garden's diversity: making good plant choices.

CHAPTER 4

CHOOSING *Plants* RESPONSIBLY

FOR MOST NONGARDENERS, plants are easy to overlook. Turn on a nature documentary and you are sure to see lots and lots of animals running and flying and hunting. Baby animals. Old animals. And the plants? Well . . . they might be in the background, but they rarely are the stars of the show. But plants are the essential foundation to virtually all other life on Earth. Without plants, there would be no cool animals to make nature documentaries about. So, before we talk through choosing and cultivating the best plants for your garden, let's take a moment to remember why they matter so much.

PHOTOSYNTHESIS

GREEN GROWING THINGS are the base of the food chain. There are a few little clusters of life in the deep sea that live off the chemical energy released by volcanic vents, and the recently discovered fungus in Chernobyl seems to be able to harvest energy from nuclear radiation. Aside from those notable exceptions, all the energy that fuels living things starts with plants and photosynthetic algae harvesting light from the sun and using it to turn water and carbon dioxide into carbohydrates. Even the fossil fuels we perhaps over-rely on are full of energy from the sun that was originally stored and captured in ancient plants via photosynthesis.

And what is true on a global scale is true in your backyard as well. Without plants, nothing else can live, so you have to start there.

REGULATING *the* CLIMATE

IN ADDITION TO PROVIDING FOOD for other living things, plants are enormous drivers of the climate of the world we live in. Plants cool the space around them by casting shade and by channeling the energy of sunlight into photosynthesis. As plants take up water through their roots and release it through their leaves, they increase humidity and cool the air. This effect can be massive. Growing plants make the climate around them cooler and wetter to the extent that tropical rainforests generate their own weather. The daily rains of a rainforest are not just due to their location on the planet—the trees themselves are releasing so much moisture that they create their own rain. On the flipslide, the destruction of plants in dry climates can set up a vicious cycle of generating drier conditions that cause more plants to die, which makes it ever drier in a process called *desertification*.

In urban environments, gardeners often have to get creative about where they grow. "Hell strips" are as good a place as any for growing native plants.

As mentioned in chapter 1, the cooling effect of plants and greenery is most easily demonstrated by what happens when they are missing—in what is called the *urban heat island effect*. Urban areas, where plants are replaced by concrete, asphalt, and buildings, are warmer than the more rural—plant-filled—areas around them, often by a significant degree, with large cities routinely 5°F to 10°F (3°C to 6°C) warmer on a summer day than the rural areas surrounding them. Now, not all of the urban heat island effect is due to a lack of plants—some of it is because our cars and buildings, especially in the winter, release a lot of heat—but it is one of the biggest factors. Adding more trees and plants to a landscape can reduce the temperature significantly, especially in the summer.

The climate-regulating power of plants plays out on the scale of your home landscape as well, with tall shade trees making your yard much more comfortable on a summer day. Positioning trees where they'll cast shade over your home in the afternoon can significantly reduce cooling costs.

On the smallest of scales, plants provide shelter from the weather for mammals, birds, and insects, most of which find places like thick grass, shrubs, hollow stems or trunks, and other nooks and crannies created by plants to protect them from heat, cold, rain, and snow.

DETOXIFICATION

Plants *want* to grow, and as they do, they filter pollutants from the air and soil.

WE HAVE PHOTOSYNTHESIS to thank for breathable air, as plants over millennia have been using sunlight to turn water and carbon dioxide into carbohydrates, a process that makes a little oxygen on the side. The plants immediately around you today don't have any real effect on atmospheric oxygen levels, as it has taken all the photosynthesizing plants and algae in the world millennia to build up the current oxygen concentration in the air, but the process of photosynthesis does allow plants to clean the air. Because every leaf needs to take in carbon dioxide and release oxygen, air has to be constantly moving through those leaves. And as the air moves, the leaves filter out pollutants, creating healthier air around them. The effect is small if we are talking about a single houseplant in a room, but outdoors, where we can have many, many plants happily growing, the impact can be enough to improve air quality, especially in urban areas.

Plants also play a huge role in detoxifying pollutants in the soil. Healthy plants produce healthy soils full of organic matter and countless bacteria, fungi, and nematodes, which can then break down and neutralize a range of pollutants in the soil. And, in many cases, even when the pollutants can't be decomposed—as with metals like lead—healthy soils can bind to the pollutants, essentially deactivating them and keeping them from leaching out to contaminate ground and surface water.

Urban plantings also offer much-needed resources for wildlife of all types.

While this collection of plants doesn't consist entirely of North American natives, it does help beautify an otherwise barren area and provide resources for pollinators.

BEAUTY

LET'S NOT FORGET one of the most valuable things that plants bring to us: beauty. Everyone can agree that lush, green, life-filled landscapes are more beautiful than barren concrete. And that impact is not some silly thing. There have been many studies showing that trees in urban areas ease stress, reduce crime levels, boost property values, and extend lifespans. Having green, growing plants sharing their beauty is necessary for us to live healthy, happy lives.

CHOOSING *Native* PLANTS

AS YOU CAN SEE, plants matter a lot. But which plants should you choose for your landscape? One of the most valuable trends of recent years is the embrace of native plants. The concept seems simple. A native plant is one that would be growing in your part of the United States naturally, before the arrival of Europeans. Native plants are usually touted for two benefits they bring to the landscape. First, they are naturally adapted to your local conditions and climate and so will thrive in your garden with less care than a plant imported from a distant land that may have evolved to thrive in quite different climatic conditions. And second, native plants will have evolved alongside native insects and other animals, so growing them is the easiest way to provide everything that your local wildlife needs to live.

This simple concept is true, but the reality of the world we live in today makes it a bit more complex.

WHAT IS NATIVE?

The first question to answer when you think about native plants is how you define the region considered "local" to you. Is a plant native to anywhere in North America to be considered a native garden plant *everywhere* in North America? Or perhaps you would consider a plant to be native only if it evolved in your general region—say, the East or West Coast, perhaps. You could get narrower still and define a plant as native only if it has been found growing wild in your state or even your county. Every nursery and gardener uses a slightly different definition of *native*, and it can be confusing to know what it means, exactly. It is generally best to be a bit flexible and use common sense when discussing which plants "count" as native. You would be generally better off focusing on selecting plants appropriate for your general climate and geographic region, rather than political lines like states or counties. If you live in mountainous western North Carolina, for example, your local climate and conditions have far more in common with West Virginia than with the coastal climates in the eastern part of your own state. Pretty much any state has a range of climates. Even within a very small geographic area, conditions can change dramatically; for example, if soil types switch from clay to sand, or if the influence of large bodies of water, be they lakes or oceans, is involved.

Really, you should think of "native" not as a simple yes/no question, but rather as a sliding scale. Indigenous plants growing wild in the woods right behind your house are extremely native. Ones found in your general region are native but a little less local, while species from quite different areas of the continent could be considered marginally native. Those from Europe or Asia or another continent are not native at all.

As we'll discuss later in this chapter, all sorts of plants, from the full range of local natives to global nonnatives, can provide value and have a place in your garden. But generally, leaning on the most local native plants possible is the best choice for your garden and your local ecosystem.

Native plants are adapted to the growing conditions in which they evolved. Choosing plants from your general climate and geographic region results in lower care requirements. For example, western species of *Penstemon* are adapted to the xeric growing conditions and soils found there.

NATIVARS

The subject of nativars—selected and named cultivars of native species—is often a source of confusion and controversy. To understand this issue, let's examine the Echinaceas (coneflowers) as an example.

Echinacea purpurea, the purple coneflower, is a species native to a wide swath of North America, with beautiful pinkish-purple blooms. If you visit a field of wild *Echinacea purpurea* and look closely at each plant, you'll see that they are all recognizable as the same species, but every individual is a little different. Some are taller, some shorter, and some might have bigger or smaller flowers, or be a slightly different shade of color.

There are so many cultivars and hybrids of coneflowers on the market these days it's tough to tell one from the other.

In addition to the natural variation within that one specific field of echinacea, there is a lot of variation over the plant's entire native range. You can find *Echinacea purpurea* in the wild from chilly Michigan to steamy Louisiana, and you can be sure the individuals at the southern end of that range are far more heat tolerant than those from the northern end, and there are almost certainly populations that have greater or lesser drought tolerance, depending on the particular climate where they grow.

Very often, when a nursery is growing a lot of a particular plant from seed, they'll notice those types of variations. They might select one that is exceptionally beautiful, with larger flowers, perhaps, or a richer color or better drought tolerance or faster growth. They might separate that plant, propagate it, and give it a name—say, for example, the selection *Echinacea purpurea* 'Ruby Star', which was selected for having large flowers and a compact growth habit. It is still a member of that species; the grower has just chosen one appealing example of the wide variation within that species and named it.

These kinds of named selections of wild species can be great because they give you a lot more information about where a particular plant came from. Any time you purchase a pot of *Echinacea purpurea* from a nursery, you are getting a particular selection of that plant. At some point, someone collected seeds from a wild population, and the descendants of those seeds are what you are buying. If the tag just says "*Echinacea purpurea*," you really have no way of knowing if that plant originally came from a population of that species that grew in your immediate area or if it came from a population on the other side of the country.

It is tempting to think that a plant with no cultivar name attached is the so-called straight species—the normal plant as it grows in the wild—somehow making it "more native" and more natural than plants that have a cultivar name attached. However, most often it just means it is a plant about which you have less information. Unless your nursery can tell you exactly where the plant originated from, "straight species" forms may or may not be native to your specific area.

However, many named cultivars are quite far removed from the species as it might appear in the wild. A named cultivar might be a particular individual selected from the wild or from a group of cultivated seedlings, but it can also be the result of intensive breeding projects. A breeder might, for example, collect the very largest-

Two of the many cultivars of echinacea on the market, one selected for its double-petaled blooms, the other selected for its short stature and orange flowers. Some of these traits don't impact the plant's ability to support wildlife while others might.

flowered individuals of *Echinacea purpurea* and start carefully breeding them with each other, selecting from each generation the biggest blooms, until they produce a plant that has larger flowers than anything you might find in the wild. That variety, again, would get a specific cultivar name, but its characteristics are far removed from what you would find in any wild population.

Plant breeding can go even further. With echinacea, plant breeders have hybridized *Echinacea purpurea* with *Echinacea paradoxa*, which has yellow flowers, combining the genes and pigments of the two species to produce a range of hybrid forms that have flowers in every shade of yellow, orange, red, and pink. These hybrid forms are the furthest removed from the wild forms of either of these species.

As with native status, it is best to think of these selected forms as on a sliding scale, with unnamed forms and simple selections from wild populations as being the closest to native, and complex hybrids as being the most removed.

Some of the changes that plant breeders make through selection and hybridization make a huge difference to wildlife, and others do not. Selections that radically change the color of flowers and leaves can make them less appealing to pollinators and insects that feed on the leaves. Some purple-leaved cultivars are found less palatable to caterpillars, for example. Double-flowered forms with extra layers of petals can make it nearly impossible for pollinators to access the nectar in the flowers (if there is any nectar produced at all), though caterpillars will still feed just fine on the leaves. These kinds of changes make a nativar act more like a nonnative plant in terms of their services to the local ecosystem, and the use of these plants is best limited in your garden. But other attributes that plant breeders select for have no negative impact on the plant's ecological value. Short, compact plants that don't need staking, plants that bloom more heavily, and plants that transplant better and are easier to grow in gardens and nurseries all are just as good for wildlife as their wild counterparts, and often are a great way to marry the needs of your local ecosystem with your desire to have a beautiful, small-space garden.

Monarda varieties bred for shorter stature and mildew resistance have perks that may make them a good fit for small-space organic gardeners. Some gardeners find these types of changes worthwhile, though other gardeners prefer to plant only local ecotypes of native plants that are grown from seed collected from nearby wild populations. Which side you stand on is up to you (even if that "side" is someplace in between), your growing conditions, and the resources you have available. Not every gardener has the means to grow only local ecotypes.

You might also shop for native plants derived from wild populations in your area. Sometimes these will be sold under specific cultivar names, but you may also find them marketed using the word "ecotype." Native plant nurseries often use this word to simply describe populations of a species from a specific area. So, you might find a nursery that advertises the sale of "local ecotypes" of a particular species, or they may describe a selection as being the ecotype of a specific area, like your county or a part of your state. Such plants might be especially well-suited to your local climate and soil conditions.

NATIVES AND CHANGING CONDITIONS

The appeal of native plants—and particularly, local forms of native species—is that they are perfectly adapted to your specific garden site because they have evolved over centuries to adapt to your precise climate and soil. This can be true, but it isn't always, because the arrival of humans in an area often represents a sudden and abrupt change from the conditions that existed before.

Climates and growing conditions have always shifted—if you live in the northern half of North America, chances are that your location has been covered with enormous sheets of ice several times in the past. But these historic changes of climate happen incredibly slowly, warming and cooling over thousands of years, allowing native species to move and adapt as conditions change.

We are all aware that, in our current moment, our global economy based on burning fossil fuels is driving change in the climate, and this change is coming at a terrifyingly rapid pace. This global-scale climate change means that perhaps the populations best suited to your garden may not be those native to your backyard, but those native to areas south of you. Migration is one of the main ways that plants and animals have adapted to changing climates in geological time, but plants migrate incredibly slowly, moving only as far as their seeds can travel each generation. There is even an ecological movement called *assisted migration* that advocates for growing native plants north of their normal range to help them move more in time with the changing climate.

In addition to the global issue of greenhouse gases and climate change, there are a lot of local factors that change the conditions in your garden as well, including the heat island effect and land development. Before your neighborhood was built, the local native plants may have been adapted to growing in a cool, shaded woodland in soil rich in organic matter. Try and grow those same plants in the degraded, sunny soil of a newly constructed subdivision and they will quickly die. So, while thinking about locally native species can be a good starting point in choosing plants for your landscape, it is important to take into consideration the actual current conditions and choose plants that are well suited to thrive there, even if those may not have been the same plants species that would have been growing on the site historically.

Opting to grow as many locally native species as possible is a good starting point for your garden.

NONNATIVE SPECIES *Can Have* VALUE

THOUGH CHOOSING NATIVE PLANTS is generally the best option, that does not mean that *only* native plants have value in your landscape and to your local ecosystem. Many nonnative plants still provide food and shelter for native insects, and they still filter pollutants, stabilize soil to prevent erosion, and provide cooling shade. Native plants are generally better, but that does not mean that nonnative plants are actively bad.

Doug Tallamy, a pioneering researcher on the importance of growing native plants in home landscapes to support native insects and birds, has led studies that have looked at native plant diversity in a landscape and how it correlates to the population of birds and insects. His research has found that a threshold of about 60 to 70 percent of the plant biomass in a landscape being from native plants provides maximum diversity for native birds and insects. In other words, your garden doesn't need to be 100 percent native plants, as long as roughly two-thirds of the plant matter in your landscape is.

There are a good number of nonnative plant species that offer value to the ecosystem as well. For example, parsley is a nonnative that provides a food source for both humans and black swallowtail butterfly caterpillars. Some gardeners opt to include a few nonnative favorites in their garden, as long as they are not invasive or aggressive. Other gardeners choose not to include any nonnative plants.

When designing and planting a garden, there may be a few reasons to keep some nonnative plants in the mix.

HARSH CONDITIONS

As we talked about, sometimes the conditions in your landscape may bear little resemblance to what they would have been before humans arrived. This means that native species just may not thrive. The most extreme example of this is street trees in urban areas. Not many species of trees can survive growing in exhaust fumes and a tiny square cut out of the concrete jungle, but urban trees provide huge benefits in air cleaning, wildlife habitat, and cooling shade. Any tree is better than no tree, so the best choice for very difficult conditions may just be a nonnative species that can handle it better than the natives can.

EXISTING MATURE PLANTS

Every landscape you will care for has existed long before you were its caretaker, and that may mean the established trees, shrubs, and perennials in your landscape are not the species you would choose. But cutting down a mature, nonnative but noninvasive tree and replacing it with a tiny native tree seedling will be a net negative for your local ecosystem for many years while that new seedling grows big enough to start providing shade and shelter. Shrubs and perennials can be replaced on a shorter schedule, but still, clearing out an established, mostly nonnative garden will be hugely disruptive to many organisms making a home there.

The best approach to enhancing an established landscape is to start by adding rather than taking away. Find open spaces and plant new native plants to complement and augment the existing landscape first. When you've filled in all the planting sites available, slowly start removing nonnative species—starting with those on invasive species lists—as you are ready to replace them with native varieties. In the case of trees, planting new native seedlings while the existing tree is still there healthily mimics how forests regenerate in nature. Every woodland has small trees waiting in the understory for a mature tree to die so one of them can take its place. If you add small native trees to your landscape, they can slowly grow and mature so that

Gardens like this one that contain a mixture of native and nonnative plants can still support wildlife, be resilient, and improve biodiversity. If you include nonnatives, be sure they are appropriate for your growing conditions and don't require a lot of external inputs (water, fertilizer, maintenance, etc.).

when your existing trees (be they native or not) reach the end of their lives and have to be cut, there are other trees ready to take their places.

SPECIFIC NEEDS

As wonderful as native plants are in their local adaptations and ability to support native wildlife, there are things they simply can't do. Food production is a huge one of these. While there are edible plants native to every part of the world, most of our modern diet is made up of an assortment of cosmopolitan species. We do need food to live, and growing nonnative fruits and vegetables locally and without pesticides is a huge ecological plus over conventionally produced food shipped to your local grocery store.

There are also many personal reasons you may want to grow a specific plant. Maybe daffodils remind you of a dear friend, or you just love big mounds of chrysanthemums in the fall. These plants can still be part of your ecologically sensitive garden, and your garden will still provide great services to your local ecosystem as long as you make sure that the bulk of your landscape is filled with native plants.

PRESERVING *Threatened* SPECIES

Our gardens can be wonderful places to grow rare or threatened plant species, such as this lady slipper orchid. However, make sure the plants are a good match for your available growing conditions and that they are sourced from ethical nurseries that do not collect plants from the wild.

ONE OF THE MAIN GOALS of having a garden full of native plants is to provide habitat and help support populations of native insects, birds, and other animals. But plants are not merely the foundation of the food chain; they are also species worth preserving in their own right. Habitat loss, climate change, and a host of other factors are threatening wild populations of native plants, and giving these species safe haven in your garden is a great way to help these species thrive and survive into the future. In fact, there are quite a few species of plants that are currently unknown in the wild and seem to have survived only because humans took them into their garden spaces. An example is the beautiful tree *Franklinia alatamaha*, which has showy white flowers and used to grow wild in coastal Georgia. The last known sighting of the tree in the wild was in 1803, but it is still grown in gardens around the world, thanks to seeds collected by John Bartram in the late 1700s and brought back to his Philadelphia garden.

Though the franklinia tree is an extreme example, and most of the plants you grow in your garden will not be extinct in the wild, your garden can be a great place to support and shore up populations of native species. In conservation terms, this is called *ex situ* conservation, preserving species in an artificial environment (like lions in a zoo), as opposed to *in situ* preservation, which is protecting species in their natural habitat (like lions in a wildlife preserve in Africa). In situ conservation is always preferable, when possible, but as more and more of our landscape gets polluted, plowed under, and otherwise damaged, ex situ conservation in gardens and other managed landscapes is a key part of preserving native species. And, unlike animals, ex situ preservation of plants can be done fairly easily, ethically, and effectively. A lion in the constrained space of a zoo will never be able to really live the full, normal life they would have in their natural habitat. A franklinia in a garden, however, lives and grows just as one would in the wild.

As you think about ways you can help preserve plants in your garden space, there are a few concepts to keep in mind.

GENETIC DIVERSITY

Even if an entire species is not at risk of extinction, there can be big risks of a loss of genetic diversity. Every species of plant or animal is made up of diverse populations over many generations, and each individual is a little different. That range of diversity is critically important to the long-term survival of a species because it is that genetic diversity that allows the species to keep evolving and adapting to survive in a changing world. Often commercial selections of native plants come from just one small population, so while there may be many plants of that species growing in gardens, they represent just a tiny slice of the true wild genetic diversity of that species. Making an effort to find diverse selections, local ecotypes, and more genetically diverse forms of familiar plants can be a great way to help preserve a healthy future for a species going forward.

When growing plants with an eye to preserving genetic diversity, it is important to be aware of the inadvertent ways we might be changing the plants through our gardening practices. Say, for example, you have permission to collect seeds from a local, wild population of echinacea. If you happen to go to collect seeds at the very end of the season, you're likely to mostly get seeds from the latest blooming individuals in that population, or even from individuals that happen to be less attractive to birds that normally eat the seeds. Ideally, you should collect seeds at the beginning, middle, and end of the time they are ripening, and make an effort to harvest randomly from a range of individuals.

Inadvertent selection can happen when growing seeds too. Most wild plants will germinate irregularly, with some seeds coming up quickly, and others taking longer. This variation is a natural insurance against a freak last frost or other weather event that can wipe out the whole population at once. Gardeners tend to want to keep the first seedlings that germinate and forget the rest, but doing so will select against the genetic diversity that helps the plant survive in the wild.

The plants in this wild collection of *Penstemon attenuates* in Idaho may all look the same to our human eyes, but they are each genetically different. The substantial genetic diversity present even in such a small population translates to increased adaptability over many generations.

Be prepared to protect plants from white-tailed deer and other threats with fencing or repellents. The overpopulation of deer in many communities means many species of wildflowers are now at risk. Shown here are two eastern woodland natives—trillium, which the deer favor, and Virginia bluebells, which they typically do not.

PROTECTING FROM THREATS

Understanding the specific issues threatening a species' health in the wild can be a great way to figure out how you can give it a safe haven in your garden. For many parts of North America, this can be as simple as building a fence to keep out deer. While white-tailed deer are a natural part of the North American ecosystem, we're seeing a rampant overpopulation of deer over much of the continent. Overpopulation of deer means overgrazing on their preferred food, devastating native stands of wildflowers and making it harder for certain species of trees to mature beyond the seedling stage. If we ever manage to re-establish natural predators for deer, that problem may be solved, but for now, protecting parts of your landscape from browsing deer can help many native plants thrive.

Fire is another big factor that has changed in many natural landscapes, especially in western North America. Before settlement by Europeans, small fires were a normal part of many of these ecosystems, and many plants evolved to require these small regular burns, either to stimulate germination of their seeds or to clear out space for them to grow. Fire suppression efforts nearly eliminated these small, regular fires, which has threatened the species that depend on them and has caused a buildup of fuel that can lead to huge, raging fires that threaten many species that could sail through a small burn.

If it is feasible in your area, getting the proper permissions to do safe, controlled, regular burns on your landscape can be a great way to create habitat for native plants. In other spaces, building a firebreak around your landscape may be a way to help protect native species from the devastating impacts of huge wildfires. See chapter 2 for more information about how to build such a fire-resistant landscape.

At other times, helping a native species thrive may be as simple as providing supplemental water through an intense, climate change–fueled drought.

These simple actions can make a huge difference in the ability of a native species to continue to thrive and grow in our gardens and in our world.

HOW *to Get* PLANTS

CHOOSING THE RIGHT SOURCE for your plants is nearly as important as the types of plants you grow. Native plants that originated right from your local area might be a terrible choice if the person you are buying them from is digging them from the wild instead of propagating them in a nursery. We never want our gardening practice to come at the expense of existing local habitats. Luckily, there are plenty of ways to propagate and grow plants without harming wild populations; it just takes a little mindfulness when shopping.

ETHICAL NURSERIES

When you are shopping for plants, look for nursery-propagated plants, rather than wild-collected plants. The vast majority of nurseries today will only sell ethical, nursery-propagated plants, but there are some unscrupulous sellers who dig plants directly out of the wild and sell them. This has become a massive problem with cactus and other succulent desert species. These species are often quite slow growing, so they are expensive for nurseries to produce, and that tempts plant poachers to dig mature plants from the wild and sell them. There are a few ways to avoid purchasing plants that have been poached in this way. A great first step is to join a local native plant group or society and get recommendations from the other members—they'll know what the good local nurseries are and may be able to warn you away from unscrupulous sellers. Next, be a little suspicious of prices that are too good to be true. Large, well-respected nurseries can give you a good sense of a baseline of what it costs to produce native plants, and if you see plants for sale for much less than that, it may well be a sign they've been ripped out of the wild.

Local garden events and markets are great places to source native plants. Just be sure they are not collected from the wild.

If an area is slated for construction and bulldozing is imminent, ask the landowner if you can conduct a "plant rescue." Always ask permission before entering the land and follow any rules they outline. Reach out to organizations that regularly conduct plant rescues in your area and volunteer.

In some cases, there are visual cues you can look for, particularly with cactus and other succulents, which are prime targets for poaching. As mentioned, a cactus in the wild will grow much more slowly than in a nursery, so wild-collected plants will typically show a much tighter growth habit, with tougher, more textured skin. When moved into a nursery, they typically start growing faster, making a clear contrast between the tight, congested spines at the bottom of the plant, and looser, more open growth at the top, the result of the new growth after the plant was removed from the wild. Finally, one of the best ways to avoid unethical nurseries is to start a conversation. Local nurseries that are passionate about native plants and ecologically sensitive gardening will be happy to talk to you about how they source their plants—and probably excited to chat with someone else who is trying to do good in the world through their gardening practices.

More and more independent local nurseries are selling native plants and cultivars of native plants, such as this coreopsis. Don't hesitate to ask questions about how plants are grown and where they are sourced.

PLANT RESCUES

Though digging plants from the wild is a terrible (and often illegal) idea, there is one circumstance where it is okay: when an area is about to be bulldozed for development. In most regions there are organizations devoted to rescuing as many native plants as possible from construction sites, and many of them will then sell those plants as fundraisers to help support their activities. Looking for groups like chapters of Wild Ones or The Nature Conservancy in your area can help you connect with people who are doing this work to either help you get involved in the rescue process or perhaps use the plants in your garden.

Often the missing link in these plant rescue efforts is the gardening know-how after the plants have been dug. One of the big reasons not to dig plants from the wild is that many of them will die if not properly cared for. Getting out with a shovel before the construction crews is just the beginning step of rescuing a plant. As you gain gardening experience, being there to help those plants get the care they need after they are dug from the wild can be critically important to helping more of them continue to live long, healthy lives.

Propagating your own plants from divisions or seeds is the easiest and least expensive way to procure more plants.

With landowner permission, collect seeds of native perennials and grow them yourself. Just don't overharvest; only take a few seeds from each plant and collect from as many different plants as you can.

PROPAGATE YOUR OWN

One of the most ecologically sensitive ways to acquire more plants for your garden is to propagate your own. When you grow your own plants, you can be certain they were not treated with insecticides during the growing process; you can be certain they were not collected from the wild; and you can rest easy that a lot of fuel wasn't used to ship them from a wholesale grower to a retail nursery hundreds of miles away.

The process of propagating plants requires some skill, and we have some basic techniques outlined in chapter 6. The simplest starting place is propagating from plants in your own garden, either by division, cuttings, or seed. There are also a lot of companies that will sell you seeds of native plants that you can then grow out on your own.

It is possible to ethically collect cuttings or seeds of plants from wild populations, but you should always do this with care. Never collect any plant materials from public lands, but if you have wild populations on your own property or have a friend who will give you permission to take propagation material from theirs, you can do that, and it can be a great way to get locally adapted native plants.

The key rule when propagating from wild populations is to do no harm, which means don't dig up any plants and don't propagate by division. You can take a few stem, leaf, or root cuttings from a plant without harming it, but limit yourself to just a few per plant, being sure to leave the majority of the plant alone. The best practice, however, is to collect seeds. Plants in the wild produce hundreds or even thousands of seeds every year, and literally just a handful of them will successfully germinate and grow into mature plants in the wild. However, with proper care from you, even just a few dozen seeds can germinate and grow into healthy plants for your garden. That means you can—again, always with permission and never from public lands—collect a handful of seeds from wild populations and grow them into mature plants for your own garden without having any negative impact on those wild populations of plants. Just like with cuttings, it is best to take only a few seeds from each plant, leaving most of them to develop normally, and collect from as many different plants as possible. This ensures you are collecting a genetically diverse group of seeds, so your garden can go on to serve as a good ex situ preservation of the wild population.

KEYSTONE SPECIES

AS YOU ARE THINKING through the many options of trees, shrubs, perennials, and annuals you could plant in your garden, consider including keystone species.

In ecology, *keystone species* are those that are the most critical to maintaining a healthy ecosystem: the ones required to keep all the other pieces of the ecosystem working properly. Species at the top of the food chain—like wolves, which regulate the populations of herbivores—are the most famous keystone species, but at the other end of the food chain there are also critically important species of plants considered ecosystem keystones.

All plants are foundational to any ecosystem, as photosynthesis is the ultimate source of energy for all the living things around us, but research has found that there are a few plants that are disproportionately important sources of food for an ecosystem. Nearly every native plant provides food to some insects, but the numbers vary dramatically. Milkweed, beloved as the only plant host for the iconic monarch butterfly, is also the host plant for about a dozen or so other butterfly and moth caterpillars, depending where exactly in North America you grow it. But a goldenrod, by comparison, can host well over a hundred different species of butterfly and moth caterpillars. Planting one goldenrod can, potentially, support nearly ten times as many different species as planting a milkweed. Some trees have an even more dramatic impact. In many parts of North America, a single oak tree can support over 400 different moths and butterflies, with wild cherries and willows coming in at around the 300 mark.

That isn't to say that milkweeds aren't important—without them, there would be no monarch butterflies, after all. But when it comes to oaks, whole ecosystems could collapse without them. Diversity in your landscape is critical to creating healthy habitat, but when you have limited space and plants to grow, choosing keystone species can dramatically increase the number of living things your landscape supports.

Milkweed (*Asclepius* spp.) is a beloved perennial that's popular among both gardeners and the dozens of insects that feed on it, including this monarch caterpillar. However, it is not the only important host plant worth including in your garden.

Polyphemus moth caterpillars are one of hundreds of different moths and butterflies supported by oaks.

When making these choices, it is important to consider not just the diversity in your yard, but the greater landscape around you. Oaks, in almost every location, top the chart when it comes to the species of insects they support, but if every house in your neighborhood already has a big old oak tree in the backyard, planting another one isn't going to make much difference. In that case, planting a maple, willow, or birch tree might be the perfect choice to increase the overall biodiversity of your neighborhood.

One great resource to help determine which plants are going to have the biggest impact in your landscape is the National Wildlife Federation native plant finder (https://nativeplantfinder.nwf.org), which allows you to search by zip code and find the plants that support the most different species in your area, or you can use a similar tool created by Doug Tallamy, which focuses on key species of trees and shrubs (https://homegrownnationalpark.org/keystone-trees-and-shrubs).

The following pages share illustrations of three keystone species found in North America and a sampling of the insects they each support.

Insect relationships for *Quercus alba*, the white oak. 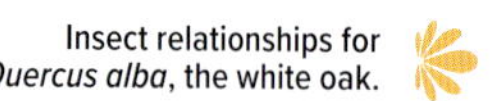

PLANT ATTRIBUTES TO LOOK FOR

ATTRIBUTE	WHY IT MATTERS	WHAT TO LOOK FOR
Novelty	Diversity of plants creates a more diverse, healthy landscape.	Choose native plants that aren't already being grown in your neighborhood.
Bloom time	Pollinators need nectar sources throughout the year.	Notice when there isn't much flowering in your garden or community and add natives that bloom then. Midsummer is often a lull.
Host plant	Caterpillars turn into butterflies or moths and are a key food source for native birds.	Choose plants that host large numbers of caterpillars, as well as specific hosts for specialist butterfly and moth species.
Fruiting/seeding	Fruits and large seeds like acorns are another critical food source for many insects and vertebrates.	Include a variety of fruiting trees and plants to support a diversity of birds, mammals, and insects, particularly during fall migration season.

Melsheimer's sack bearer
phyllonorycter fitchella
red-humped oakworm
phyllonorycter aeriferella
coptotriche zelleriella
american oak lacebug
red-winged sallow
common oak
oak leaftier
phyllonorycter albanotella
scalloped sack bearer
fourteen-spotted leaf beetle
waved sphinx
girlfriend underwing
the slowpoke
curve-lined looper
white-m hairstreak
pink-striped oakworm
yellow-winged oak leafroller
hedgehog gall
spiny oakworm
oak beauty
spiny oak-slug
oak besma
oak leaf phylloxera
solitary oak leafminer
distinct quaker
oak leaftier
oak skeletonizer
Schlaeger's fruitworm
orange-tipped oakworm
afflicted dagger
angulose prominent
variable oakleaf
ilia underwing
oak blotch leafminer
American dun-bar
oak webworm
sleepy duskywing
fawn sallow
buck
large maple spanworm
green-dusted zale

Insect relationships for *Helianthus annus*, the sunflower.

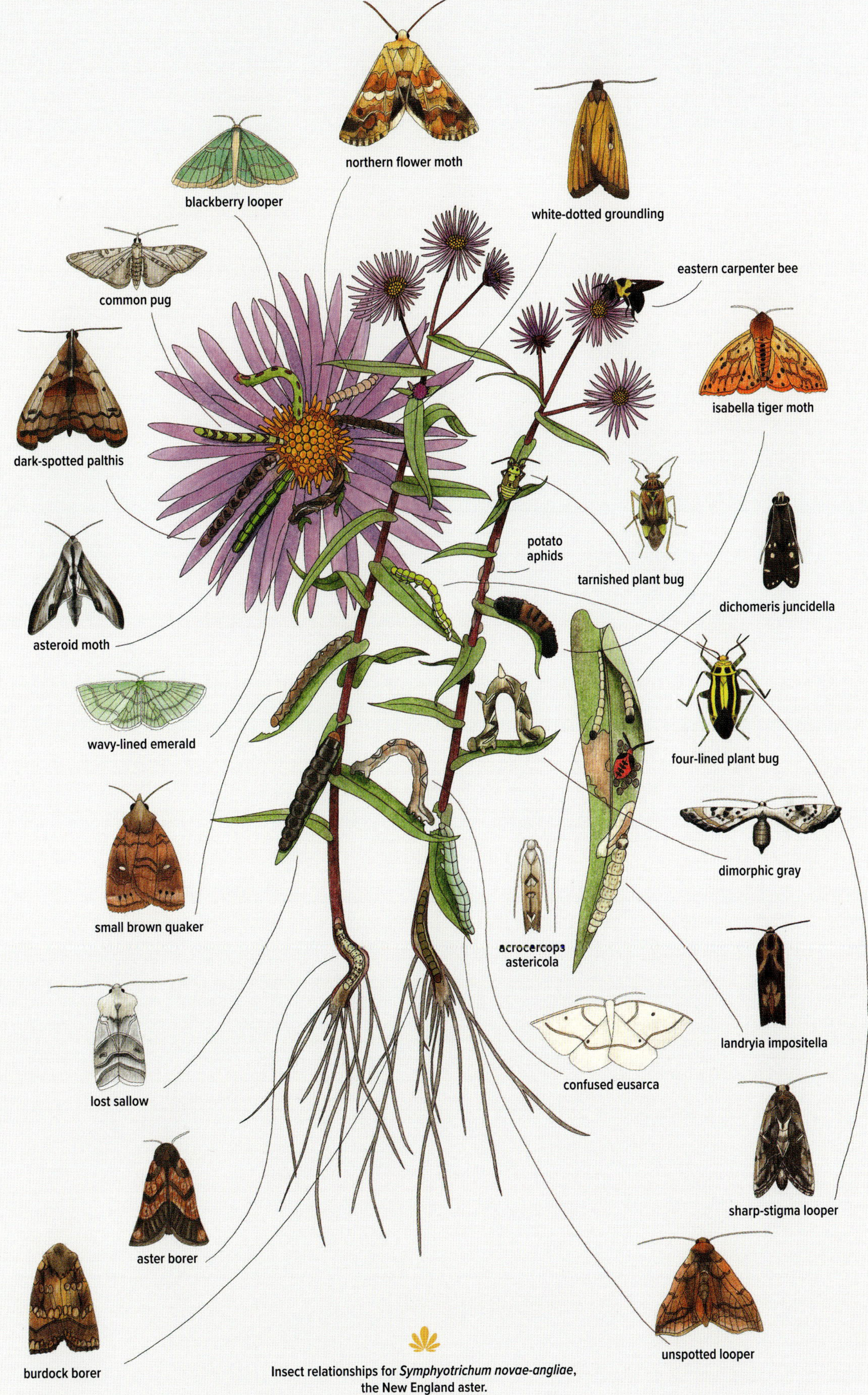

Insect relationships for *Symphyotrichum novae-angliae*,
the New England aster.

LAYERS *and* FORM

IN ADDITION TO CHOOSING plants that are powerful in their ability to support a range of species of birds and insects, it is important to think about how each plant we add fits into the larger landscape. A great way to do this is to think in terms of layers, ranging from tall trees to smaller trees, shrubs, perennials, and ground covers. Different birds, insects, and other living things make their homes in different layers, so including all of these layers in the landscape is a great way to maximize the diversity your garden can support. Thinking in terms of plant layers is also a win-win in terms of garden design. Richly layered gardens are just as beautiful to look at as they are vital for a healthy ecosystem.

One caveat in terms of layers in the garden is that the average number of layers found in your particular climate will depend largely on your rainfall. Rainier climates, like those in eastern North American and in the Pacific Northwest, are dominated naturally by very tall canopy trees. As rainfall amounts drop off, the tall forests shift to scrublands, where the tallest layers are shrubs and small trees. Eventually this shifts to grasslands—first tall-grass prairies and then short-grass prairies as the average rainfall amounts keep dropping off. In dry climates, the taller layers will only be found along rivers and other pockets of higher moisture, and supporting tall shade trees in a dry climate will likely require a significant investment in irrigation that is often not sustainable long term. Re-creating the number of layers that are natural to your particular climate will require the least amount of work and be the most sustainable. It also re-creates the habitat that the wildlife in your area has adapted to thrive in.

Ecologically diverse landscapes have many layers and include a combination of both woody plants and herbaceous ones. They include many bloom shapes, colors, and times.

Don't neglect the ground cover layer. Finding native species for this layer can be a challenge, but it is worth the effort for the reduction in weeds and watering. This gardener has combined sedges with white wood asters for a thick ground cover in a shaded site.

GROUND COVER

The ground cover layer is the shortest part of the garden: low-growing plants that cling to the ground. These low-growing plants can be a huge boon to the garden because they act like a living mulch, preventing erosion, reducing water loss, and even enriching the soil over time when they drop their leaves and stems as they grow. Another huge plus for gardeners: ground covers also can reduce or nearly eliminate weed problems, as they can effectively shade out weed seedlings before they get established.

In a mature garden, much of the ground cover layer may be tucked away out of sight, growing in the shade of a bigger plant or even filling in a background area that isn't easily seen, and they often don't have the big, exciting flowers of taller perennials and shrubs, so many gardeners are tempted to skip planting them. But don't be fooled: Ground covers are key, behind-the-scenes supporting players that make everything else in the garden run more smoothly.

Unfortunately, if you go to many nurseries and look for the ground cover section, you'll probably see a lot of very aggressive, rapidly spreading, nonnative species for sale, including notorious invasives like English ivy and common vinca. The rapid spread of these species makes them great at quickly covering the ground, but unfortunately, they can be overly vigorous and crowd out other plants. Often the best species for a ground cover layer are not the traditional, rapidly spreading species, but short, clumping perennials that can be planted thickly in open areas.

The real rock stars of the ground cover layer are sedges—members of the genus *Carex*. Sedges look like grasses, though botanically they are not a true grass, and there are many native species suitable for nearly any gardening conditions and most regions. They can be planted thickly to mimic the look of a traditional lawn, and many of them are quite shade tolerant, so they can fill in difficult areas under shade trees. One of the really great uses of native sedges is interplanting them as a ground cover layer under taller perennials. Many native perennials, like baptisias, have showy flowers and beautiful foliage, but also an open, vase-shaped habit that leaves a lot of bare ground at their feet. A shade-tolerant sedge can be planted around them to thrive under the perennial canopy and fill in those open areas. You might not even notice an effective sedge ground cover layer in a full, mature, perennial garden, but it'll be doing a great job of supporting even more living things and protecting the soil. Even better, many sedges are evergreen, or nearly so, so they keep the garden looking full and green even when the bigger perennials are dormant for the winter.

In arid climates, you'll see more bare ground in natural habitats, and that can carry over into your garden, too, but the judicious use of ground-covering plants can be very helpful in these climates as well. Many small native cactus, for example, tend to grow in the sheltering shade of larger plants, and can be planted as a beautiful and useful ground cover layer in even the driest of landscapes.

PERENNIAL

Perennials are one step up from the ground cover layer, but this layer can also include small shrubs and annuals. This is the layer that is taller than the ground covers, topping out around 3 or 4 feet (0.9 or 1.2 m) tall. This is perhaps the easiest layer to overdo in the garden, just because there are so many really beautiful, showy, flowering plants that fall into this category.

Showstopping coreopsis, phlox, monardas, goldenrods, and asters all cover themselves with flowers, and let's be honest: Most gardeners are suckers for flowers. Lots of flowers feed lots of pollinators and predators, but do remember that you can't just rely on this layer alone in the landscape. Tucking less sexy ground covers into the open spaces and making room for bigger shrubs and trees is critical for creating a diverse, well-rounded landscape that is a great home for a range of bird and insects. Aesthetically, it will look better as well. As pretty as flowers are, often they look better when grown in clumps that contrast with other sizes and types of plants.

Your plant palette will look different depending on where you live. This garden in the desert southwest focuses on a collection of cacti, succulents, and local native perennials like penstemon.

The perennial layer is often the most colorful with an incredible diversity of native plants available to home gardeners.

SHRUB

The shrub layer is made up of small, woody plants. The distinction between a shrub and a small tree is a blurry one, but generally shrubs have multiple stems, rather than a single trunk. The shrub layer brings a lot of huge benefits to the garden. Generally, shrubs have dense, twiggy forms that are critical safe havens where birds and other living things hide from predators. Many birds like to feed in more open, sunny areas, but they don't like to venture too far from a dense shrub where they can dive for cover. Because the woody stems stay up all winter, even in the coldest climates where the leaves may fall off, shrubs are an important place for birds, mammals, and insects to shelter in the winter, where they can find a little protection from chilly winds.

As a practical matter, shrubs are a huge boon to the busy gardener because they are incredibly low maintenance. Their permanent woody structure allows them to quickly shade out any potential competing weeds, and there are no dead stems to decide whether to cut back each fall as there are with an herbaceous perennial. While you may be tempted to rake the leaves from a tall shade tree in the fall, the lower growth habit of a shrub means fallen leaves are mostly out of sight, still serving to enrich the soil and provide lots of habitat without offending even the tidiest gardener's sense of aesthetics.

The biggest challenge with shrubs is in the planning. As woody plants, they will continue to get bigger and bigger each year, and that means a cute little shrub in a 1 gallon (3.8 L) pot you buy today may become an unwieldy monster as the years go by. So, when purchasing a shrub and deciding where to plant it, look up the mature size, and then actually measure out those dimensions before putting it in the ground. Planting a shrub that will get 4 feet (1 m) wide 1 foot (30 cm) away from your front door is going to be an endless headache.

When shopping for dwarf conifers, which are incredibly useful in the landscape visually and provide safe places for animals to shelter during the winter, be particularly aware that they will keep growing their entire lives. The mature size listed on most tags will just be how big you can expect them to be after five to ten years. But twenty years from now, they'll be twice that size. Give them room to mature, or ensure you have a plan to keep them pruned as they grow.

Native shrubs, both evergreen and flowering, fill a unique niche in the ecological garden. This buttonbush (*Cephalanthus occidentails*) supports endless bees and butterflies and thrives in moist to wet soils, including in rain gardens.

Planting tiny shrubs at the correct spacing can leave a garden looking very bare and empty for a few years, so you should plan to add short-term plantings between them. These can be annuals, shorter-lived perennials (steer clear of very long-lived varieties), or you can even put other shrubs in those spaces, choosing cheaper varieties you're willing to prune back and eventually remove as their more desirable neighbors fill in. The goal here is to plant with a plan so the garden will look good when freshly planted and ten or twenty years from now.

UNDERSTORY TREES

Small trees like redbuds, dogwoods, Texas mountain laurels, desert willows, and magnolias are commonly found in the wild at the edges of woodlands; growing underneath taller canopy forest trees; or filling in as the tallest layer in dry scrublands.

In an urban landscape, understory trees may be the tallest you can feature for practical reasons. Tall canopy shade trees, as beautiful and marvelous as they are, are not practical to plant under power lines or in very small lots. If the utility company is going to have to resort to extreme pruning to keep a tall tree from hanging over power lines, it is generally a better choice to plant a small tree to fill in the space. Small trees are incredibly useful in the landscape. They provide shade and nesting habitat for birds, plus many of them flower beautifully in the spring. There are also many small fruit trees that provide a delicious harvest for you to eat after the spring flowers have faded. Few plants can provide the ornamental value, edible harvest, and ecological benefits of a small fruit tree.

Because small trees grow up with a single trunk, unlike shrubs, they also open lots of opportunities for gardening underneath them using shade-tolerant perennials and ground covers. Remember that the line between shrub and small tree is very blurry, and you can push plants one way or the other with careful pruning. If you cut out all but one or two of the stems of a large shrub and prune away some of the lower branches, you can encourage it to grow into a small tree form to open up gardening space beneath it. Similarly, some small trees can be pruned down low to encourage multiple trunks and keep them in a shrub form, depending on the requirements of your garden space.

CANOPY TREES

The big trees—oaks, maples, firs, and the like—are the tall trees of a mature forest canopy. These large trees can take a lifetime to mature, so they have to be managed a little differently in the landscape. The other landscape layers fill in relatively quickly, totally transforming a space. While larger trees may grow large enough to offer some shade in a fairly short time, it may take decades for them to provide maximum shade. That is why the first priority in a landscape is to preserve and maintain existing large trees. Providing supplemental irrigation during extreme droughts can be helpful, and if pruning is needed to keep a tree away from power lines or for other practical reasons, be sure to hire a licensed, qualified arborist to do it properly and in a way that will not harm the tree. But your top priority should be protecting tree roots. The majority of tree roots are in the top layer of the soil, and extend a huge distance out from the tree, often a far wider reach even than the width of the branches. Minimizing digging in the root zone will help a tree thrive. Never let cars drive or park over the soil around the base of a tree as this will compact the soil and can have a long-term impact on the health of the tree.

The other important consideration with large trees is to plan for their replacement. In a forest, the tall trees in the canopy always have small, young trees waiting under them. You can mimic that process in your own landscape, planting small trees in the spaces between big mature ones so that when they do eventually reach the end of their lives, it won't take so long to replace the tree canopy.

The benefits of tall trees in the landscape are many. For most of eastern North America, tall forests are the dominant native ecosystem, so many of our native birds and insects have adapted to thrive in them, and many species of tall forest trees qualify as keystone species, supporting an incredible number of caterpillars and other living things.

In addition, tall trees completely transform the landscape around them, cooling the ground with their shade and significantly slowing down wind speeds to create a much more sheltered habitat.

Tall forest trees also have strong relationships with mycorrhizal fungi in the soil. We've talked already about how these fungal networks help make plants more drought resistant and uptake nutrients better. You can support these networks in the ground by providing lots of organic matter and keeping the soil mulched. Plugging a healthy tree into that network makes it even stronger.

The biggest challenge to growing tall canopy trees in many landscapes is their sheer size and their potential to do damage when they blow down or lose branches during storms. A tall tree leaning over your home may provide valuable shade in the heat of the summer, but no one wants a tree falling on their house. When planting trees, keep this in mind, and choose locations where they can grow to a mature size without the risk of damage. You can still, for example, benefit from the shade of a tree if it is strategically placed to the south or west of a house to cast shade over the home during the hottest parts of the day. And choosing to keep open, sunny areas up closer to buildings or vehicles that could be damaged by trees is very effective. If you have a large tree that is a potential hazard, your best course of action is to consult a qualified and properly licensed arborist to evaluate the health of the tree and the potential for damage.

Small understory trees, like this alternate leaved dogwood (*Cornus alternifolia*), offer habitat and food for birds and bring structure to even the smallest garden.

Large canopy trees, like those in the backdrop of this garden, offer many benefits, particularly in eastern North America, where forests were the dominant native ecosystem before the arrival of Europeans.

Nature's Companions in the Vegetable Garden

Companion planting is the science, and art, of growing two or more plants near one another so they can support each other. You may also hear this discussed as intercropping or plant associations. Maybe you know of the Three Sisters planting method. This Indigenous planting tradition puts corn, beans, and squash in the same garden plot. The corn grows tall, providing a support for pole beans to climb. The beans are legumes, so nodules on their roots work with soil bacteria to fix nitrogen in the soil. The squash grows large leaves that shade the ground and smother weeds. The Three Sisters is a great illustration of how we can leverage plants' natural characteristics to support one another. There are a number of reasons you could consider using companion planting in the ecological vegetable garden. We'll look at those here.

ATTRACTING INSECTS

In the ecological garden, you'll select plants based on their many ecosystem services. Attracting the right bugs is an important one. Pollinators are a priority category of insects, but you have the option of working with many others. (Find a chart of beneficial insects on page 81 for some of the best bugs to know.) You can help these good guys find your garden with companion planting.

Combine flowering annual and perennial plants with vegetables in edible gardens to reap many different benefits, including attracting pollinators and improving the soil.

One ornamental-edible companion to know is coreopsis interplanted with cabbage. With its beautiful flowers, coreopsis attracts hoverflies, and hoverfly larvae are aphid predators. Aphids, meanwhile, feed on cabbage. When the adult hoverflies visit the coreopsis flowers, they'll lay eggs nearby, and when the eggs hatch, the larvae can have their first meal of aphids on the cabbage.

Pollination is an obvious companion-planting benefit, especially in gardens that are mostly fruits and vegetables, which rely on pollinators for production. Pollinators will be attracted to the flowers put out by these fruits and vegetables, of course, but the more nectar and pollen sources they have to bring them into a space, the better. Look at the size and shape of the flowers that your fruits and vegetables are producing and learn about the time of year these flowers are blooming. Use companion plants that mimic these characteristics to bring in the insects that feed on them. Keep something blooming all year long so the pollinators are always active.

DISTRACTING INSECTS

Vegetable gardeners might plant a crop as a trap crop to spare their more valuable vegetables. An example is planting collard greens near cabbage. Diamondback moths love cabbage but not as much as they love collards. While some moths will eat the cabbage, most of them will go after their preferred crop. In this companion-planting situation, the gardener may not need to turn to pesticides to control the diamondback moths because fewer of them will bother the cabbage.

In your ecological vegetable garden, you can plant a trap crop, wait for it to become infested with the pest, and remove the buggy plants from the garden. This garden waste goes in the trash rather than your home compost so the pests leave your property. Plan your trap crops to mature ahead of the main crop, and don't plant them adjacent to one another, lest the pests find the plants you're trying to protect. How far apart to plant your trap crop from the desired one depends on which pest you are targeting. If the pest is small and not very mobile (say, aphids, for example), the trap crop should be just a few feet away. But if the pest is highly mobile (say, adult squash vine borers), then several dozen feet would be better.

REPELLING INSECTS

Just as companion plants can attract the right insects, they can also repel the wrong ones. Nasturtium and marigolds, for instance, can help reduce damage in your garden from cucumber beetles and squash bugs.

As you explore companion planting for insects' sake, look for research-based advice. Don't be afraid to do your own research comparing plants in your own garden, but don't spend a lot of money and time on unproven plant combinations. (The information in this book comes from science-backed sources.)

Legumes like peas and beans convert nitrogen from the air into a form usable by plants. They make great plant partners in the vegetable garden.

IMPROVING SOIL

Leguminous plants are usually the first to come to mind for plants that improve soils, particularly in vegetable gardens. Their nitrogen-fixing capacity is important in the edible garden, and it's helpful that there are so many lovely legume plants to choose from, such as beans, purple prairie clover, and climbing peas.

Simply having live plant roots in the soil is a means of supporting the soil web of life and building soil fertility. This is a benefit of having a living mulch, a ground cover of low-growing plants. Many ground covers serve other companion-plant services as well, like attracting pollinators and suppressing weeds.

Other companion plants that provide soil fertility are those with taproots that break up heavy clay soils and mine minerals from the depths. The plants include carrots, marsh mallow, and even alfalfa.

WEED SUPPRESSION

Any plant that covers wide swaths of ground helps to suppress weeds' ability to grow, including the large leaves of the three sisters' squash. Fast-growing ground covers will also keep weeds from germinating.

CLIMBING SUPPORT

To add depth and interest to your space and give your vining plants a place to sprawl, you could erect arbors and trellises in the garden. From a companion planting perspective, tall, sturdy plants serve the same purpose. Corn, sunflowers, and okra are all tall garden plants that can double as living trellises, as they won't shade out their nearby companions.

There are few more wholesome-sounding gardening ideas than companion planting. Companion planting as well as planting to attract, distract, and repel certain insects are examples of how to leverage nature in ecological gardening.

SEASONALITY

Come spring, queen bumblebees and carpenter bees emerge and head straight for the blooms of blueberries and other spring-flowering shrubs and perennials.

PLANTING ALL THE LAYERS of the landscape fills up every possible niche in space, but you can also think about layering the garden through time, ensuring that you have growing, blooming, and fruiting things through each of the seasons to support your local wildlife throughout the year.

SPRING

In climates with cold winters, spring is a critical time for native insects. Many of our native pollinating insects hibernate over the winter, and when they emerge in the spring, they need to feed to carry on through the new season. Bumblebees, in particular, need early spring flowers to survive. Bumblebees make small hives, but unlike honey bees, the whole hive won't necessarily survive the winter. The workers all die off, and only a few new queens hibernate through the cold and emerge in the spring. When they first emerge from hibernation, the queen bumblebees have no workers, and so they have to establish the new hive, lay their first eggs, and feed their first batch of workers all by themselves. Once the first workers have emerged, the queen can settle down to just lay eggs while the workers collect food for the hive. But in the earliest days of spring, she's on her own. Early spring-blooming wildflowers native to your area are probably the most critical food for these bumblebee queens. Growing these flowers in your garden will support these native pollinators, many of which have declining populations and need all the support they can get. Be aware that many early spring bloomers are ephemerals—meaning they go dormant shortly after flowering—so plan to interplant them with perennials that emerge later to fill in the spot over the summer.

For birds and many mammals, spring is also baby season. Eggs are hatching in nests and cute little fuzzy animals of all sorts are taking their first adorable steps out into the world. Survival of these young animals depends on a reliable food supply, of course, which your landscape will be providing in the form of leaves to eat and insects to dine on, but these creatures also need safe places to hide. This is where having dense shrubs, and other thick, somewhat overgrown areas is critical. Wide-open spaces—which we often prefer aesthetically—are very unsafe for baby animals that aren't yet able to run, hop, or fly fast enough to escape predators.

In summer, accessible water is essential in the ecological garden. Keep water features and birdbaths clean and filled.

SUMMER

As we move into the warmer parts of the year, one of the key features for local wildlife is the availability of water. During hot, dry spells, putting out bowls of water for wildlife is much appreciated. Be sure to include some shallow dishes of water with pebbles placed on them, which gives insects a place to rest while they drink. Do so during a drought and you'll be surprised how many bees, butterflies, and other insects you'll see stopping by to hydrate.

Because fewer plants bloom in the height of summer than in spring and early summer, it can be a challenging time for pollinators—particularly those in hot climates. If, during the dog days of August, your garden is mostly green and without many flowers, this is a good time to visit a local nursery to find summer-flowering plants to add to the garden.

Summer is also when shade is the most valuable to wildlife, so take time to walk around and see where a tall tree might be welcome to cast some shade over a water source for wildlife, a patio for you to sit on, or your home to reduce your need for air conditioning. Shade trees are a long-term project, so while they are maturing, you can use temporary shading from umbrellas or shade sails to help cool your garden space.

Come fall, floral resources are especially important to fuel overwintering and migrating insects. And don't forget to include lots of native grasses, which serve as excellent habitat when left standing all winter.

Dozens of bird species feast on seeds in the garden. Plus, these plants bring plenty of beautiful winter interest to the landscape.

AUTUMN

The fall is when many animals require the most food, either as fuel to migrate to warmer climes or to fatten up to hibernate through the winter. Prioritizing planting keystone trees like oaks that produce huge quantities of nutritious acorns will be a huge boon to your local wildlife, and including lots of trees and shrubs with colorful berries in the fall will help your local bird population migrate or prepare to live through the winter.

It is critical not to forget to keep feeding pollinators through the fall as well. Some, like monarchs, will be undertaking epic migrations; others will hibernate through the winter; and still others will be stashing food away in tiny caches with the eggs of the next generation. Fall bloomers are sometimes easy to overlook in the garden. Most of us only shop for plants for our gardens in the spring. It is way too easy to get caught up buying all the plants covered with spring flowers and overlook all the perennials that *will* look incredible and be flower-covered in the fall, but are just boring green lumps in the spring. Prioritizing adding great fall-blooming plants like asters and goldenrods to your garden will provide a critical food source to your local pollinators and keep your garden looking beautiful for as long as possible. Be sure when you are out plant shopping in the spring to bring with you a list of the fall-blooming species you want to add to your garden and buy them even though they might not look like much at the moment.

WINTER

Winter can look dramatically different in your landscape depending on your climate. In the cold, snowy parts of North America, it is a time of dormancy. The best thing you can do for wildlife is to ensure you leave them plenty of safe places to hibernate in the form of piles of leaves and brush, and lots of dense, twiggy shrubs and perennials that aren't cut back in the fall. And for the animals not hibernating, like many birds, you'll want to supply food through the winter in the form of seed heads on perennials and lots of plants loaded with berries for them to snack on through the cold months.

In warmer climates, there will be more going on in the winter, and even more mouths to feed, as many birds and some insects from the north will migrate down to more southern winter gardens to wait out the snowy months before heading back north in the spring. That means in climates with mild winters, you want to prioritize having a great food supply available during the winter months—lots of flowers, insects, seeds, and berries.

CHOOSING *for* RESILIENCE

AS WE CHOOSE PLANTS for each of the four seasons, we're faced with the reality that each of those seasons can throw unexpected curveballs our way. Try to think back to a year that you would describe as "normal" in the garden—has there ever been one? Whether it is a summer that is shockingly hot, a winter that is terrifyingly cold, rains that just won't stop, or a punishing drought, there is always something unexpected to deal with. Plus, the normal unpredictability of the weather now has the added layer of a changing and destabilized climate that has already warmed significantly and will continue to do so.

While the general impact of greenhouse gases is to warm the climate, the specific, local impacts can be variable and hard to predict. While warmer temperatures and increased frequency of droughts are likely outcomes for many regions, rainfall patterns could change to produce significantly wetter conditions in some areas. There is evidence suggesting that a warming climate may increase the chances of a destabilized polar vortex, resulting in more extreme cold snaps in winter.

In addition to weather and a changing climate, there are other threats our gardens may face. New pests and diseases are a continuing challenge. Our global economy ships goods all over the world and has allowed new fungi and insect species to arrive on new shores, with sometimes devastating impacts on our plants. From diseases like Dutch elm disease and chestnut blight to insects like the emerald ash borer and spotted lantern fly, these problems are an unfortunate reality of the connected world we live in today.

While we can't know exactly what challenges our gardens will be facing next month, let alone ten years from now, we can choose plants for our gardens in ways that will make them more resilient and better able to withstand whatever the future holds.

As our gardens face more pressure from extreme weather events and introduced pests and diseases, we need to do what we can to mitigate the potential negative effects by focusing on native plants and encouraging diversity and resilience.

DIVERSITY

The power of having a diverse landscape made up of a range of different plants has been emphasized many times in this book, and it is absolutely key to building a resilient landscape. A simple example of how powerful a diverse planting scheme is, is the case of the American elm and street tree plantings. Wherever you live in North America, chances are there is an Elm Street near you, which, when it was first built, was planted with a line of elm trees on each side of it. American elms were hugely popular street trees for their beautiful vase-shaped habit, rapid growth, and overall vigor and health. Elm trees thrived with ease, so they were planted widely throughout our cities and towns, and not just on the streets named after them. Then, in the early 1900s, logs of European elm wood were brought to the United States and accidentally introduced two closely related species of the sac fungi that cause Dutch elm disease, and suddenly the perfect street tree started dying.

Some streets were lined exclusively with elm trees, but the disease transformed these beautiful, shaded stretches into a totally open, barren space with no tree cover at all. Facing this huge loss, people planted new trees. A top choice was ash trees, which are fast growing with beautiful fall color and relatively free of pests and disease . . . until, that is, the early 2000s, when a beetle called the emerald ash borer was spotted in Michigan and proceeded to kill nearly every ash tree in eastern North America.

These twin invasions devastated some streets twice; their elm tree cover was lost to disease, only to be replaced by rows of ash trees that then fell to the beetles.

A Texas meadow filled with blanket flowers is also filled with more genetic diversity than you may think. Each plant is slightly different, and over time, the growing conditions select for the traits that matter most to their survival.

But other street tree plantings fared far better. Instead of planting just one species of tree down an entire street or all through a neighborhood, some areas used a mix of different trees. These included elms and ashes, but also maples, oaks, and poplars. In the face of these new challenges, the neighborhoods with diverse tree plantings lost a lot of trees, but they didn't lose their tree cover—the other species thrived and kept the streets shaded and beautiful as the dead trees were cut down and replaced.

There is no way anyone could have predicted either Dutch elm disease or emerald ash borer, and we don't know what the next invasive pest or disease will be, but we can learn from these stories by not putting all our eggs in one basket. The more diverse our plantings, the better able they are to recover from the loss of any one particular species.

The resilient power of diversity in the landscape doesn't just apply to new pest and diseases—extreme weather events, droughts, floods, cold snaps, and heat waves all impact different plant species differently. Including a range of different species in your garden ensures that even if one dies, your garden is still full of thriving plants giving beauty and creating habitat.

DIVERSITY WITHIN A SPECIES: THE POWER OF SEEDS

In addition to hedging against unexpected future challenges by growing a range of different species in your landscape, you can also select for diversity within each species by choosing seed-grown plants over clonal selections. As we talked about earlier when discussing nativars and other selections of wild plants, every species is made up of many genetically diverse individuals. When plants are grown from seed, each plant is genetically a little different. But clonal propagation—via cuttings or division or tissue culture—creates many genetically identical plants. That can be a good thing—many named selections of plants are particular clones that have exceptionally good characteristics, like vigor, heavy flowering, or particularly intense color.

It is also useful, sometimes, to have all the plants of a species be genetically identical, particularly if you are creating a very formal garden design where small variations of height and color would throw off the perfect symmetry you are trying to achieve. But in the face of a changing climate or other challenges, that genetic uniformity can be a weakness. In a genetically diverse population, some individuals may be more heat tolerant or better able to resist a new disease, so chances that some individuals will thrive through whatever the climate and weather and pests throw at them are much better.

Even better than just growing a genetically diverse population is to regularly save seeds from the plants in your garden and grow them again next year. Saving your own seed is particularly practical with annuals—if you love zinnias, for example, you can simply save seeds from your favorite plants each year and not have to spend money on new seed packets. But the wonderful thing about that is not just that you save money, but that you allow the species to continue to evolve in response to a changing climate. If you grow zinnias from seed each year, and each year save seeds from the healthiest and prettiest individuals, you will be allowing them to continue to evolve and adapt to changing climate conditions. If an intense drought wipes out most of your plants, but a few survive, save seeds from those tough survivors and your next generation will be more resilient when the next drought comes.

During droughts, make note of which plants are more resilient and include more of them in your future planting plans. You can also save seeds of the more drought-tolerant plants to select for that trait in subsequent generations.

EMBRACING THE LEARNING OPPORTUNITIES

When gardening disaster strikes, and extreme weather beats down or kills your plants, it is easy to get frustrated and angry. But you can also take these moments as a chance to learn more about your garden and build more resilience by simply looking at which plants thrived despite the harsh conditions. If you have a richly diverse landscape, you will have some species and even individuals within a species that will thrive through all but the worst climatic emergencies. And even in conditions that don't out and out kill plants, there will be some that struggle and look ugly, and others that continue looking perky and beautiful no matter how hot, cold, wet, or dry it gets. When the garden is struggling the most, go out and make note of which plants are still thriving. Make a list, or even just snap a photo with your phone as a quick way to record the best plants. Then, when the bad weather is over and you are replacing plants that died, consider replacing them with more of the plants that thrived.

If, for example, an extreme heat wave seriously damages half the plants in your garden and you replace them with the plants that did great despite the heat, well, the next time there's a similar heat wave, your entire garden will sail through unscathed. Extreme weather events are also a great time to walk around the neighborhood and visit public gardens to see which plants are still doing well. Consider adding these stalwarts to your own landscape. Approaching it this way can ensure your garden grows more and more resilient with each passing year.

PLANTS MATTER

PLANTS ARE THE FOUNDATION of the food chain and the source of food and shelter for nearly every other living thing on this planet. With proper plant choices, you can build a garden that supports an abundant and vibrant ecosystem, is durable and resilient in the face of changing climates and extreme weather, and, perhaps most importantly of all, will bring joy and beauty to your life.

CHAPTER 5

WATER STEWARDSHIP

ONE OF THE TRUISMS about gardeners is that we always complain about the weather. One year it rains too much and we're fighting waterlogged soil, erosion, and even flooding. The next, it stops raining altogether and we are staring down the brutal reality of drought. But it doesn't have to be that way. There is a Scandinavian saying that "There is no such thing as bad weather, just bad clothing," and just as good, warm clothing is the solution to cold winter weather, good landscape design is the solution to the vagaries of annual rainfall. Yes, the weather will always be unpredictable, and there are occasional intense, devastating weather events, but by and large, thoughtful design and water management can create an ecologically vibrant landscape that is resilient and will thrive through the normal variations of rainfall.

Smart water management is also the right thing to do for the environment. In dry climates, regularly irrigating large swaths of the landscape is simply unsustainable and uses up a shocking amount of our limited water resources. According to the Environmental Protection Agency, nearly a third of residential water use is devoted to irrigating lawns and gardens, adding up to an average of 9 billion gallons (34 billion liters) of water every single day. Smart choices can reduce that water use dramatically.

But just as limiting irrigation needs in dry climates is critically important, it is equally as important to design landscapes that handle water smoothly when it does rain, particularly when it rains too much. Torrential rains can cause flash flooding, erode precious topsoil, and send pollutants to contaminate rivers, lakes, and other waterways. Pollutants—especially fertilizers—carried with runoff are the primary cause behind devastating "dead zones" in places like Lake Erie, the Chesapeake Bay, and the waters off the coasts of Oregon and Washington, where nutrients flowing into the water cause huge algae blooms that then die, sucking all the oxygen out of the water and killing the fish. In the Gulf of Mexico, these dead zones appear each summer, often growing to the size of the state of New Jersey (or the country of Wales).

Well-designed landscapes can capture water, slow it down, and allow it to soak slowly into the earth. The simple process of letting water soak into the ground recharges groundwater supplies (the very thing we're sucking dry to keep irrigating our lawns), reduces the chance of flooding, prevents erosion, and significantly reduces the pollutants that run off into our streams, rivers, lakes, and oceans.

As with most things in an ecologically centered landscape, there may be some tradeoffs you'll have to make to create a more water-wise landscape, but there are also many ways in which the ecologically sensitive choice is also going to save you money, time, and create a more beautiful landscape.

MANAGING RUNOFF

Rather than letting rainwater run off your property and down into storm sewers, an important tenet of ecologically responsible gardening is to find a way to either capture it for reuse or slow its movement and encourage it to soak into the ground on-site.

WHEN IT RAINS, no matter whether you live in a wet or dry climate, the goal should always be for rainwater to soak down into the soil on-site, rather than rapidly running off the surface of soil or pavement and into storm drains and waterways. Slow absorption into the soil reduces the chances of damaging flooding, but even more important is the filtering action of soil. Rain arrives as nearly pure water, but the moment it hits the earth, it starts absorbing everything it touches: spilled gasoline from a lawnmower, trash on the side of the road, freshly spread fertilizer, or even a nutrient-rich pile of excrement left by a neighborhood dog. Soil can filter out some pollutants and nutrients as the water filters down through it, preventing these potential contaminants from running off or leaching into waterways and wells. But if that contaminated water runs straight down a storm drain and into a river, *all* of those pollutants run right with it, where they can have devastating consequences, particularly in dead zones created by fertilizer pollution.

Water-wise landscapes can be quite beautiful. Their main goals are to reduce runoff and lower irrigation needs.

The other benefit to slowing and capturing rainwater runoff is to build resilience for dry periods in the future. Water that soaks deep into the soil remains there, available for plant roots to take up, and, eventually, it can soak in deeply enough to start recharging the groundwater sources we tap with wells during dry seasons.

The slogan for managing runoff is: "Slow it, spread it, sink it." Anywhere you have concentrated, fast-moving water on your property, you want to find ways to slow it down, spread it out, and allow it to soak down into the soil.

Along with the ecological imperative of slow it, spread it, sink it, we might add another, more practical one: Move it. Specifically, move it away from your home. Heavy water flow around the foundation of your house and other buildings can damage foundations and flood basements, so, whenever possible, direct water away from buildings to other areas of the landscape where it can soak into the soil without causing any harm.

MAP YOUR WATER

The first step to managing and improving the flow of water on your property is to figure out what the current situation is. The easiest way to do this is to wait for a heavy rainfall, get an umbrella or a good rain jacket, and go outside to see what is happening. You might, for example, notice that one of your downspouts empties onto your driveway, where the water then flows down to the street and straight into a storm drain. Elsewhere, you might find water is pooling up right by your foundation because the slope of your yard is pointing back toward the house. A hill in the backyard might have rain rushing down it quickly, heading straight into a small creek, carrying debris, fertilizers, and pollutants with it.

With these observations in mind, sketch a quick drawing of your property and use arrows to show where the water is flowing and collecting. Anywhere the water is leaving your property or rushing quickly over the soil surface is a chance for you to redirect it, slow it down, and allow it to soak into the soil. Areas where water pools after a rain can be altered to help the water soak in faster, or if pooling happens somewhere unfortunate, like the middle of a pathway or next to your foundation, the water can be redirected to collect somewhere else.

PERMEABLE SURFACES

Our landscape is made up of two types of surfaces: permeable ones, like soil or gravel paths, that allow water to soak down through them, and impermeable ones, like concrete and houses, which shed water. Impermeable surfaces make runoff problems worse, so limit their use when possible. For example, an elevated wooden deck is a more permeable choice than a concrete patio, and there are even options for permeable concrete for driveways and sidewalks that allow water to soak through them. Some impermeable surfaces are inevitable—a permeable roof on

Dry creek beds are one way to divert rainwater away from your home's foundation and allow it to soak into the ground farther away.

This illustration is an example of how water can be directed using a swale with berms. Water is channeled over the rocks, where erosion is limited. Plant material at the sides further prevents erosion and the berms keep the water funneled toward the swale.

your house, for example, would be a huge problem—but when you have the option, choose a porous surface over one that will shed water.

SWALES AND BERMS

The simplest way to direct water where you want it to go is by building swales and berms. A swale is a small, usually grassy or gravel-covered depression—like a drainage ditch but shallower—that encourages water to flow along it. A berm is the inverse of that—a low mound of soil that directs water away from it. You could use a swale to direct water from a downspout away from your house toward somewhere it can soak safely into the ground. Building small berms perpendicular to a slope can stop water from rushing straight down a hill, instead slowly moving from berm to berm. You might also build a berm along a sidewalk, driveway, or other impermeable surface to stop water from flowing off into a storm drain.

Swales and berms do not need to be large. Raising or lowering the ground just a few inches can direct water to flow where you want it and can be quite subtle in the landscape. Alternatively, you can make them bigger and more dramatic to add interest to a flat yard and create opportunities to grow different plants that prefer the drier tops of berms or the wetter bottoms of swales.

A newly installed rain garden situated in a residential yard. Notice the incoming pipe and the gravel-lined depression to allow the water to soak in slowly.

RAIN GARDENS

A rain garden is a sunken area of ground designed to capture rainwater and allow it to sink slowly into the ground. Usually, rain gardens are made by digging out a depression, building swales to guide water into the low area, and then adding a berm on the downhill side to keep the water from flowing out of the area. In the bottom of the rain garden, a mix of very light sandy soil is added to help the water soak into the soil. The presence of the light soil is the difference between digging a rain garden and a pond. On the swale side, it is critically important to build a spot for overflow. After a very heavy rain event, if your rain garden fills completely up, the water will start flowing over the berm. If it runs over the soil, it will quickly erode the berm, and you'll have no more rain garden. To prevent this issue, build a spillway, slightly lower than the rest of the berm and lined with gravel or even concrete. When the rain garden fills up, excess water flows out of the spillway and the gravel or concrete lining stops the water from eroding the berm.

Rain gardens consist of a depression to capture rainwater from downspouts and allow it to slowly sink into the ground. The bottom is a sandy or gravelly soil. Plants that tolerate occasional "wet feet" are the best candidates for rain gardens.

Rain gardens can be built anywhere that water collects, but they are generally designed specifically to catch runoff from the roof of a building and other impermeable surfaces in the landscape. A good rule of thumb is for a rain garden to be about 10 percent of the size of the impermeable surfaces that will be draining toward it.

This front yard water garden in an urban neighborhood is filled with plants that tolerate periodic flooding. It is an oasis for pollinators and other wildlife.

EROSION CONTROL

WHENEVER WATER MOVES across your landscape, erosion occurs—the water eats away at the soil and carries it off as sediment. Reducing erosion is critically important both for the impact it has on your landscape and for the impact it has downstream.

At home in your landscape, erosion can cut unsightly gullies through your property, and when it's especially bad, it can even make slopes and structures unstable. This is important every day but becomes critical during intense rain events and floods where erosion-prone slopes can transform into devastating landslides. Even when erosion is not so dramatically destructive, it is very harmful to your landscape because it reduces the nutrient-rich topsoil your plants need to thrive.

The other reason preventing erosion is so important is because all of that eroded soil has to go somewhere. Soil particles can clog up storm drains, cloud the water of creeks and rivers, and all of the rich nutrients in your topsoil that could have been supporting a lush garden end up instead polluting streams, lakes, and oceans.

Slopes prone to erosion are best covered in dense plantings, ideally of regionally native plants.

In vegetable gardens, the presence of cover crops prevents erosion and soil loss when the beds are fallow.

KEEP IT COVERED

The simple first step to preventing erosion is to keep the ground covered. Bare soil is quickly eroded away by wind and rain. Mulches can be used to reduce erosion, but the real key to erosion-proof soil is thick and healthy plant life. Plants don't just cover the soil; their thick network of roots lock the soil in place. In fact, healthy, plant-covered soil can actually grow thicker and deeper with time, rather than eroding away.

If you have bare, erosion-prone soil, your first priority should be to get it covered with plant life as soon as possible. And if you already have a plant-covered surface, you should be very careful about removing those plants. This applies even if the plants in question are a noxious weed or an invasive species. While generally removing invasive species from a landscape is a good thing, if they are the only thing holding the soil in place, removing them all at once can be devastating, especially on a slope. If you have plant material you want to get rid of growing on an erosion-prone slope, focus on replacement rather than removal. Clear just enough of the invasive species from the area to plant a native one, and then slowly cut back and remove the invasive as the native plants expand.

For example, if you have a slope covered with invasive Japanese honeysuckle shrubs, rather than ripping them all out at once, prune back the branches enough to let light get to any native shrubs or trees you have planted. Keep slowly pruning the honeysuckle back each year as your native plants get established, and only fully dig them out when the native canopy is thriving. This may seem like a lot of work over a long period of time, but it is much easier than trying to keep a slope in place while a whole new population of plants gets established. Once the topsoil has eroded away, there is no getting it back, and many of your new plantings may fail anyway.

The soil most frequently left bare is that devoted to growing food, with most crops being grown as annuals in the summer months and nothing being grown over the winter. Because vegetable gardens do best with fertile soil, it is particularly important to prevent that soil from eroding away, both to keep the fertility high and to avoid polluting waterways. There are a few simple things you can do to help reduce the potential for erosion during the offseason. First, after your tomatoes or peppers are killed by frost in the fall, instead of pulling them out, which rips their roots out of the ground and disrupts the soil, simply cut them off at ground level. This way, the roots will still be there to add stability to the soil as they slowly decompose over the winter. Next, be sure to add some kind of mulch to cover any bare soil.

Straw and shredded fall leaves are great options for a vegetable garden. You can also add this throughout the normal growing season as well if there is bare ground between your plants. Finally, consider growing through the winter months to keep the soil locked into place with roots. Depending on your climate, you may be able to grow cold-tolerant crops like kale, lettuce, and other hardy greens through the winter months. And no matter where you garden, you can grow cover crops. Cover crops are fast-growing plants sown any time the ground is bare to lock the soil into place with their roots. They also add organic matter as they decompose, increasing soil fertility, and they can reduce pest and disease problems thanks to the crop rotation break from the usual food crops.

Though the principles of cover cropping are usually deployed for the vegetable garden, they can be just as useful for annual flower gardens. If you grow petunias along your sidewalk for the summer or a row of zinnias for cut flowers, don't leave the soil bare in the winter.

Traditional cover crops can be deployed in a cutting garden, but there are plenty of more ornamental plants that can be used the same way. Winter annuals like pansies are a great choice in all but the coldest climates, and spring bulbs like traditional daffodils or crocuses are a terrific option as well. Though spring bulbs don't show growth above ground until the spring, they start growing roots in the fall, helping to keep soil locked in place during the winter months until it is time to plant your annuals again in the spring.

MANAGING SLOPES

The topography of your landscape influences how big of a problem erosion will be. While level ground, if left bare, will erode away from wind and water, the problem is much bigger on slopes, and the steeper the slope, the more critical it is. There are a few actions you can take to minimize erosion on slopes.

The most dramatic option is terracing, which involves building a series of walls perpendicular to the slope, effectively transforming the slope into a set of "stairs" with level ground between them. Terraces lock the soil in place and create level areas you can garden intensively on without having to worry too much about erosion. They are, however, expensive, and should be installed by a professional. The walls of terracing, especially on steep slopes, must be carefully and precisely engineered to ensure their long-term stability. They must also contain places for water to drain through them without undermining the walls. Improperly installed terraces can fail suddenly and dramatically, forcing a whole hillside worth of erosion to happen in a few moments.

Because of their expense, terracing is best suited to areas where you want to garden intensively. If you have no level ground, but want a vegetable garden, terraces are your best option. If, on the other hand, you have a mix of topography on your property, it is probably best to keep the frequently disturbed soil for vegetable and annual gardens on a piece of level ground and grow perennials, trees, and shrubs on any slopes to keep the soil locked in place permanently with roots.

Stopping shy of full-on terraces, there are other ways to help stabilize slopes. Short-term solutions can keep the ground from moving while new plantings get established.

Terraces can be created in sloped areas by building walls perpendicular to the slope. This gardener has used rocks to build their terraces and filled each resulting terrace layer with a mixture of perennials, grasses, and low-growing evergreen shrubs.

One of the simplest methods is to hammer stakes into the ground and then lay fallen branches or logs horizontally across the face of the slope, the stakes serving to hold them in place. The branches will slow the flow of water, catch some of the soil, leaves, and other debris moving down the slope, and stabilize the ground above them for planting trees, shrubs, and perennials. The branches will eventually break down, but by then, the new plantings will be established and can take over the job of holding everything in place.

No matter how carefully you stabilize and plant a slope, there will be some erosion, so it is a good idea to add berms or a rain garden at the base of a slope, if you can, to catch the water, sediment, and nutrients coming off the slope and stop them before they pollute waterways.

FLOODING

Flooding can be very problematic, especially in low-lying areas. Build flood tolerance into your landscape by considering the addition of spillways, barriers, swales, and berms to channel excess rainwater to safe places.

AS IMPORTANT AS DESIGNING a landscape that will be resilient during droughts is designing one that will stand strong when a flood comes. Weather extremes are a fact of nature, and as much as we hope we'll never have to face them, it is always better to be prepared.

FLOOD RISK

The first step to getting your landscape prepared for a flood is to understand the flood risks in your area. Flood risk maps are available from the Federal Emergency Management Agency, and you can search them for your address and see what flood risks are associated with the place where you live. Generally, low-lying areas near rivers and oceans are most at risk, as well as narrow valleys in mountainous regions that can quickly concentrate rainwater. Flood risk assessments are based on historical weather and flooding data, as well as an analysis of the topography. But even the best flood risk estimates are just estimates, and as the climate continues to change, we should all expect the unexpected and be prepared for floods that were never a problem historically.

SPILLWAYS

During an intense rain event, water coming down from the sky onto your property can fill up all your rain gardens and containment systems and then . . . what? All the plans to capture and conserve water with berms and barrels need to also involve a plan for what happens when those fill up. So, any structure you build to catch water should have some kind of spillway built into it to direct excess water away before it builds up to a level where it can threaten your house. In rural areas, spillways can be directed to drainage ditches, streams, and rivers. In urban areas, the target is city storm drains. Wherever the water is headed, it is important to regularly inspect the course and clear out any leaves, branches, or silt that has built up and may stop the water flow. Many localized urban floods have been caused by leaves clogging the entry to a storm drain.

BARRIERS

During very serious flooding events, the threat will not just be the water falling on your property, but it may also come from rising river or ocean waters that enter your property from somewhere else. In the event of a predicted flood, you can transform your water-catching berms into property-protecting dikes simply by putting sandbags over spillways and extending the heights of any berms to try and keep water away from buildings.

AFTER THE FLOOD

If your garden has been inundated with floodwaters, recovery needs to be done with care. Easily visible after a flood is the acute damage: the raised beds that have washed away, the paths that have been covered with silt, and the plants that have been damaged or killed. But far less visible are contaminants in the floodwaters. Raw sewage often gets mixed into floodwaters, and there is a chance that heavy metals—like lead—or other chemical contaminants have been carried with the floodwater as well. Avoid tracking possibly contaminated soil and dust into the house by removing dirty boots and gloves before entering and washing thoroughly. Do not harvest food from a flooded vegetable garden for 90 to 120 days after the event, particularly crops like greens that are eaten raw. Adding compost and a thick layer of mulch will bind many chemical contaminants and reduce their chances of getting on food or clothing.

After a flood is a good time to make further plans to create a more resilient garden going forward. If you can see where floodwaters entered, consider building a berm to fend off future floods, or invest in taller raised beds to keep the vegetable garden up out of floodwaters. We can never control the weather or eliminate all damage from storms, but learning from each disaster we face can make our landscapes ever more durable.

The good news is that there is an upside to excessive rain. With a rain catchment system in place, you'll be able to harvest plenty of rainwater for future use.

PREVENTING *Wasted* WATER

Rain barrels are useful for capturing rainwater for later use. Be sure there is a screen in place to filter out debris and keep mosquitoes at bay. Models with built-in faucets are very handy.

THERE ARE MANY WAYS to conserve, capture, and reuse water to keep it on-site.

GRAY WATER

Sewage, of course, needs to be properly processed to avoid causing pollution or making people sick, but a lot of the water that heads down our drains is what is called *gray water*—it's not suitable for drinking, but it's also not polluted with sewage. In many cases and with the proper precautions and permits, gray water can be safely used to irrigate ornamental gardens. This includes the water from our showers, sinks, and dishwashers. A gray-water system can be professionally installed to capture this water and direct it out to the landscape for irrigation. But do check on local regulations before installing a system like this. In areas with very low water resources, your municipal water system relies on recapturing gray water to recharge, and so it is not permitted to direct it out for irrigation. In other areas, doing so is just fine, and certainly it is a great idea if you have a well or are or otherwise not connected to a municipal wastewater system.

Short of installing a full-on gray water capture system, there are lots of easy ways you can reuse water that would otherwise go down the drain. Washing vegetables from the garden? Instead of doing it in the sink, do it with a hose over a bucket outside, then pour the water on the vegetable garden when you are done. Boiling some potatoes? Set your colander over a bucket when you drain them and save the water for the garden. Do not save water that has harsh cleansers in it, or any water from cooking that has a lot of salt added to it, like most pasta-cooking water.

RAINWATER HARVESTING

Another big source of wasted water is your downspouts. You can direct that water to rain gardens to allow it to soak into the ground, but you can also capture it in rain barrels or cisterns to use for irrigation. Like gray water, there may be municipal codes forbidding this practice in the driest of areas, as cities want to ensure that rainwater soaks into the ground to recharge ground water, so check with local regulations before installing a rainwater-capture system.

A rain barrel is the simplest rainwater capturing vessel—just place a large container under your downspout and direct the water into it. It can be as simple as that, but there are a few steps that make them function better. First, always put a fine screen over the top of the rain barrel—it will let the water in while screening out leaves and other debris, but most critically, it will stop mosquitoes from getting into the barrel to breed. Second, you need to plan for overflow. Install a second downspout that comes out from the top of the barrel so that once it fills up, the excess rain is still directed away from your home's foundation. Finally, elevate the rain barrel on some bricks, stones, or cinder blocks, and install a faucet at the bottom. As long as your rain barrel is higher than the rest of your landscape, you can rely on gravity to send water from the barrel out your hose and into the garden. You can purchase ready-made rain barrels, or, with a few DIY skills, build one yourself.

A rain barrel won't hold much water and won't likely provide enough for irrigation needs if you live in a dry climate. To really get serious about harvesting and capturing rainwater, cisterns are the way to go. These are very large containers that are often installed underground. Underground cisterns, of course, can't rely on gravity to move water, so you'll need to install a pump. Because of this, cisterns are much more expensive to install than rain barrels, but the payoff is enormous as they can harvest and store large quantities of water to get you through dry spells.

Cisterns are larger vessels for rainwater capture. They can be above ground, as shown here, or below ground, depending on your preference and local ordinances. Underground models may require a pump to move the collected water to where it is needed.

IRRIGATION

Rain chains are an attractive alternative to downspouts. This one leads to a below-ground cistern, where the water is stored for later use.

THE SIMPLE ACT OF ADDING WATER to plants is the most basic of gardening tasks, but it can also be the most ecologically fraught. In the drier parts of North America, the desire for irrigation water has led to the damming of rivers to make reservoirs. So much water is being sucked away that the Colorado River barely makes it to the sea any more, and pumping out of underground aquifers in the San Joaquin Valley in California has caused ground levels to drop by nearly 30 feet (9 m) in some areas.

In general, ecologically sensitive gardening means limiting the amount of irrigation water you use. We've talked a lot in other chapters about ways to manage your soil and choose the right plants to create a resilient landscape that can thrive without requiring a lot of supplemental water. There are still plenty of reasons why you'll want to add water to your landscape, and many ways to make that process more efficient and effective.

There are a few basic principles to keep in mind any time you are irrigating plants. We'll talk about how to implement these principles in the sections that follow. Keeping these in mind will go a long way to maximizing irrigation efficiency.

Drip irrigation systems and soaker hoses allow water to penetrate the soil slowly, reducing runoff and evaporation and targeting the water directly at plant roots.

WATER MORE DEEPLY LESS OFTEN

When irrigating, it is tempting to grab a hose or run a sprinkler until the soil looks wet and then stop. The process is then repeated the next time the soil looks dry. This practice, however, irrigates only the very top layer of the soil, while deeper soil layers stay bone dry. This is a problem because moisture from the soil surface is the most prone to evaporation, and because plant roots grow where the water is. Watering just the top of the ground results in plants with shallow root systems that suffer at the first sign of drought. Instead, when you water, add enough to soak down deep into the soil, where the water will be protected from the drying sun and wind. This encourages plants to grow deep, healthy root systems.

You can test how deeply you've watered by waiting an hour or so after irrigation stops (to give the moisture time to seep in) and then digging a small hole with a trowel. If only the top couple of inches of soil are wet, you need to water for much longer.

Once you've done a deep watering, you won't need to irrigate again for a while. Even if the soil surface looks dry, again, grab a trowel and see what it looks like deeper down. Some dry soil—1 to 2 inches (2.5 to 5 cm)—at the surface is just fine as long as there is moisture deeper in the ground.

WATER SLOWLY

Just as with reducing erosion and runoff, when irrigating, slowing the water down is the best way to make sure it soaks in. Turning your hose on full blast and pointing it at a dry plant will permit very little water to soak down deep; most water will simply run off, often eroding the soil with it. This is particularly critical, of course, on slopes. It can be easiest to achieve slow-but-steady watering with an automatic irrigation system, but even if you are watering with a hose by hand, you can slow down the rate of water delivery by watering once, taking a break to let that soak in, and then coming back to add more water a moment or two later.

Ollas (handmade, like this one, or commercial models) are buried vessels that are filled with water that slowly soaks out into the surrounding soil. They are made of porous clay and can be used in containers or in the ground.

WATER ONLY WHERE NEEDED

This may seem obvious, but watering things that don't need irrigation—like driveways or patios—is a huge source of wasted irrigation water. Poorly aimed sprinklers waste a lot of water, and slow leaks where hoses connect to faucets can lose a surprising amount of water, deeply saturating one spot of ground beneath the leak. Even if you are directly adding water to plants, it doesn't mean it is needed. Many plants can survive and even thrive through far more dry conditions than we give them credit for.

MINIMIZE EVAPORATION

Once water hits the soil, we'd like it to stay there, rather than dissipating as water vapor. Deep watering, as discussed above, helps with that, as does keeping soil well mulched. Timing can matter a lot here as well. Water added in the evening will soak deeper into the soil before it faces the evaporating heat and sunshine of the following afternoon. Sprinkler design can make a big difference here too. Sprinklers that throw water high in the air allow a lot of moisture to evaporate away even before the water hits the ground; the problem is particularly bad if the water comes out as a fine mist rather than larger water droplets.

IRRIGATING NEW PLANTINGS

Many of the most drought-resistant plants—especially those native to North American prairies—thrive through dry spells by growing incredibly deep root systems capable of supplying the leaves with water even when the soil surface is quite dry. But, of course, after you pull a plant out of a 1-gallon (3.8 L) pot and stick it in the ground, it takes some time for such a deep, drought-resistant root system to develop. As a result, you will need to add extra water for the first year—or possibly two—after planting, as the roots get established. This is a situation when it is very important to water deeply. Your goal is to encourage deep roots, so be sure to water long and slowly enough to get moisture deep into the soil.

For smaller plantings, the best way to do this is by hand with a hose. For larger areas, it may be worth setting up a sprinkler or a drip irrigation system. If you have planted trees, a great solution is a product like a gator bag, which is filled with water that drains out slowly through small holes in the bottom of the bag, allowing the moisture to soak slowly and deeply into the ground. This is an incredibly efficient way to make sure new tree plantings get the moisture they need to thrive with almost no waste.

DROUGHT

WATERING DURING DROUGHT is fundamentally tricky. The moment when we most want to add water to our landscape is when there is the least water available to give our plants. During a drought, we should be doing our best to save water, not use more of it, and yet, we don't want our plants to die, nor do we want the insects and birds that rely on those plants to starve. What is a gardener to do?

The first step to navigating irrigation during a severe drought is to decide which plants to prioritize. Generally, you should focus on saving long-lived, hard-to-replace plants—especially trees—over fast-growing plants like annuals. If you lose your vegetable garden or a bed full of annual flowers due to drought, you can just replant them the next year, once the drought breaks, and your garden will look exactly the same. Let a mature shade tree die in a drought, however, and it will take decades to repair that damage. Longer lived perennials fall somewhere in between. You'll also want to prioritize plants that are special to you. Common rose varieties, for example, are pretty fast growing and you can easily buy new ones. But an heirloom rose variety gifted to you by a friend or family member, on other hand, is an irreplaceable treasure, so give it the water it needs.

Drought can mean different things in different places. The official designations of drought produced by the United States Drought Monitor are based on deviations from the norm. This means that the same amount of rain might qualify as a severe drought in the usually rainy southeastern North America or an unusually rainy season in the desert Southwest. In the garden, what we want to know is whether or not it is dry enough for our plants to start suffering. Depending on your local climate and plant

Drought-tolerant landscapes are essential, especially in areas with low rainfall. This dry climate garden contains a collection of primarily native plants with a few drought-resistant non-natives.

choices, you could be in official drought conditions, but your plants may still be thriving. If you live in a normally rainy climate, you may not pay much attention to whether a plant is listed as drought tolerant or not, but when the rain stops, it is a good time to pull out your favorite references and see what they say about how much drought a particular plant can take. One good way to get an idea is to look for recommendations for climates much drier than yours. If you garden on the East Coast and a plant is recommended by gardeners in Arizona or Colorado, it will probably sail through even the worst drought.

You can also look at plants and get an idea of how drought tolerant they might be. Silvery, fuzzy, succulent, or waxy leaves are all adaptations plants have evolved to thrive in dry conditions, while thin, delicate, bright green leaves are more typical of plants adapted to wet conditions.

Tree bags (sometimes called gators) are placed around newly planted trees and regularly filled with water. The bag is perforated at the bottom, so the water slowly drips out to irrigate the tree with little to no water wasted.

When you do decide to water during a drought, be sure to water deeply. It is much better to give your beloved tree one good deep soaking during a dry spell than to sprinkle a little water over the soil surface every few days. Somewhat counterintuitively, this can mean the best time to water is right before or after a rainfall, if you happen to get one. A brief rain will not do more than wet the soil surface, but adding more water afterward can help water soak deep into the ground where it can do a lot of good. Watering before rain can be helpful in a drought, too, as sometimes extremely dry soil becomes hydrophobic, meaning that water will just run off the soil surface rather than soaking in. Moderate irrigation will moisten that extremely dry surface layer and help rainwater soak in.

Droughts are the best time to use clever water-saving tricks, like saving gray water. No matter the presence of drought, you'll be showering and washing vegetables, so save that water and put it on the garden rather than sending it down the drain.

Just as your plants need moisture during a drought, so do all the other things that live in your garden. Keeping a few bowls of water around the garden—regularly changed out to keep mosquitoes from breeding—will allow the birds, mammals, insects, and other wildlife to keep hydrated during a dry spell. You'll be surprised how many critters you'll see visiting every day.

The final step to navigating a drought in the landscape is to learn from your experience. During a dry spell, one of the best things you can do is keep careful records of how long the drought lasted, what you watered and what you didn't, and how your plants performed. And be sure to keep a close eye on how things recover once the drought ceases as well. Peonies, for example, may lose all their leaves and appear to die during a dry spell, but in fact they have just gone dormant and will reappear the next spring. Learning how things respond to a dry spell gives you critical information on how to react to the next drought. And if some plants did die, consider replacing them with more of the varieties that did not. The practice makes your garden more resilient and easier to care for, ensuring that during the next dry spell you can relax instead of running around with hoses and worrying.

IRRIGATION SYSTEMS

THE MOST ECOLOGICALLY SENSITIVE, sustainable, and low-work version of a landscape will be filled primarily with plants that won't need supplemental water once they are established. However, sometimes gardeners want to build regular irrigation into the garden. Food production is a huge reason—even if you live in a dry climate, growing your own vegetables is an ecological plus over the commercial vegetables you buy at the store. Yes, you'll use water to grow them, but so will commercial growers. At home you can be thoughtful about avoiding pesticides, and you won't need to burn fossil fuels to ship vegetables across the country. But vegetables aren't the only reason to irrigate your landscape. In a fire-prone areas you may want to keep a well-watered firebreak around your home to protect it (see page 54). And sometimes there is a treasured plant that is worth a little extra water for personal reasons, like a memorial planting established in the memory of a loved one or pet, or heirloom plants passed on by friends or family.

Building an irrigation system into your garden's design can be useful, especially if you live in a dry climate. Just be sure the system sends water directly to plant roots and is maintained regularly to prevent leaks.

You may also be able to deploy regular irrigation in an entirely sustainable way depending on where the water comes from. If your water is coming from wells, pumped up from aquifers, or being siphoned off of rivers and other natural waterways, the less you use the better. But if you are able to harvest and store sufficient water in rain barrels and cisterns, you can use that water to irrigate in a completely sustainable fashion. This is quite possible in many climates on the West Coast that are rainy during the winter and dry during the summer. If you invest in water-harvesting capabilities, you should be able to capture enough water during the rainy season to use during the dry periods.

IRRIGATION ZONES

The first step to planning an irrigation system is to divide your property into irrigation zones. Designate areas where you will provide regular water—perhaps a vegetable garden or highly visible areas with showy flowers near the house or in outdoor seating areas. Next, designate occasional water areas, places that will be fine without supplemental water most of the time, but you'll want to be able to water occasionally during a dry spell. This might be a beloved shade tree or flower garden that needs a little help to get through the heat of the summer. And, finally, designate the no-water areas, places where plants thrive on simply what falls naturally from the sky. Irrigation zones are critical to efficient irrigation. If you scatter plants willy-nilly, it is almost impossible to provide water to all the plants that need it without also watering the plants that are just fine on their own. Hydrozoning and keeping the regular water areas small and massed together allows you to easily keep them healthy while not wasting water where it isn't needed.

When designating irrigation zones, you might also want to look back at the traits you discovered about your property when planning to control runoff. Areas where runoff naturally collects—like the bottom of slopes and the edges of impervious surfaces like your driveway—will already be the wettest areas on your property, so if you position your regular irrigation zones there, you'll be able to grow more water-loving plants while using less irrigation water.

CHOOSING A SYSTEM

There are a million and one types of irrigation systems, but they all are made up of hoses that carry water from your faucet (or cistern or water barrels) to emitters that release the water to your plants.

The hoses of your irrigation system are easily overlooked but can be a prime source of water loss. They can be buried underground or laid on the soil surface, and in either case, you should know where they are and how long the material is expected to last. Buried hoses in particular can develop leaks and waste a huge amount of water without you even realizing it. Know where the hoses are and be prepared to replace them as needed.

Emitters range from traditional sprinklers that spray water over a large area to tiny drip-emitters that release water very slowly through small holes. They all work, but there are some rules of thumb to keep in mind when choosing the most efficient system. Generally, emitters that spray or drop water a short distance are more efficient than those that spray it high into the air. Ideally, choose emitters that release the water under the canopy of your plants so it can soak straight into the soil rather than hitting leaves, where it can promote disease problems and be lost to evaporation. Also look for "pressure-regulated" systems when possible.

Ideally, irrigation systems should be set up in zones with high-water plants kept separate from low-water plants. This allows you to run irrigation more frequently to the plants that require it and avoid irrigating plants that don't want or need it.

A cheap sprinkler runs on whatever water pressure comes from your faucet. Water pressure that is too high will lead it to overspray its intended targets and waste water, and very high water pressure can cause misting—fine sprays of water droplets so small that they easily blow away in the wind and evaporate rather than soaking into the soil. Water pressure that is too low, on the other hand, will mean not enough water makes it to all the intended targets. Pressure-regulated systems independently keep water pressure constant, so you can rest assured that everything is getting what it needs and no water is being wasted.

Finally, you'll want to consider a control system. The simplest—and cheapest—option is just a faucet you turn on when you want the irrigation system to run and turn off when you are finished. This works as long as you aren't a forgetful person and don't spend a lot of time traveling. It can be tedious, and it is easy to get distracted and let sprinklers run much longer than needed. Simple timer systems that turn the water on and off at a set time and day are the next step up. They are not very expensive, require less attention, and make it easier to water during the evening and night when there is less evaporation. Finally, you can get smart controllers that monitor rainfall, temperature, humidity, and even weather forecasts to deliver water exactly when it is needed. Some water utilities even offer rebates to those who install smart controllers. These systems are costly, but they can be worth the investment by making irrigation stress-free and efficient.

MONITORING THE SYSTEM

Whatever irrigation system you set up, it is critically important to regularly check it and make sure it is functioning properly. Particularly if your irrigation system runs at night, it can be very easy to overlook a broken emitter that is no longer spraying properly or a leak in one of the hoses. Every month or two, turn everything on when you are home and walk around to make sure all the spray patterns are functioning as intended. Next, turn everything off, check your water meter, and then check back again in 30 minutes or so to make sure it hasn't moved. If all your hoses are off and your water meter is still moving, you have a leak that needs your attention.

There are many types of irrigation systems. Small, ground-level emitters, like the one shown here, deliver water directly to the base of individual plants.

Using basic automatic control systems to schedule irrigation lets you water while you're away and manage how long your system runs. There are also smart controllers that monitor rainfall, temperature, and other factors to deliver water in precise amounts only if it's needed.

Final THOUGHTS

WATER RIGHTS AND MANAGEMENT are already huge issues and a political hot button in many parts of North America, and as climates change and populations grow, they are only likely to become bigger issues in more areas. Extreme weather events—whether they be droughts or deluges—are incredibly stressful, difficult times for gardeners and wildlife. We can never eliminate all of the stress of managing water and weather extremes, but taking the time to plan ahead and design a thoughtful landscape that handles weather extremes—both wet and dry—allows you to relax and spend less time worrying and more time enjoying the landscape and garden around you.

Water is essential to all the life in your garden, humans included. Being smart about its use and preservation is key to ecological gardening. 

CHAPTER 6

HANDS-ON *in the* GARDEN

WITH A SOLID UNDERSTANDING of the thoughts and practices that go into ecological gardening, it's now time to get to work. Techniques used in the ecological garden may be similar to those used in a conventional garden, though you're always thinking about how your work will affect the larger natural community. One thing that makes gardening interesting is that there are different ways to achieve the same goal. Rather than offer quick-fix garden hacks, this chapter looks at how to take care of your garden tasks with the full ecosystem in mind, starting with installing a new garden, various maintenance tasks, and plant propagation basics.

BREAKING *Ground*

PROPER GARDEN PLOT preparation is an important part of setting up a successful garden. As with other gardening techniques, you have options for how to prepare a new planting area, and these options have their own ecological and logistical considerations. The path you choose will depend on the tools and resources available to you and how the method aligns with your ecological gardening principles. It will also depend on the plants you plan to include in your garden and the style of garden you're creating. Vegetable garden prep will be different from preparing a site for a meadow planting because vegetables need more organic-matter-rich soils than do native meadow plantings. Whichever method below you choose based on these factors, start with closely mowed grass with the clippings left in place.

SHEET MULCHING

The most labor-intensive method of converting a lawn to a garden is also the most ecologically minded. Sheet mulching involves layering cardboard, compost, and mulch on top of the ground to smother grass while building a base for the garden to come. You might hear this called *lasagna gardening,* because the garden is built in layers like a lasagna.

The first layer, cardboard, is a ubiquitous resource. Visit any home-appliance store to be wowed by the amount of refrigerator- and dryer-sized cardboard sheets they'll allow you to take home. Any size of cardboard will do, but the big-box sheets are especially nice for covering large areas. Layer the pieces so no spot of ground is uncovered; it only takes a speck of light for grasses and weeds to push up and out. Remove the labels, tape, and staples from the cardboard before placing it.

Layer compost and mulch 3 to 6 inches (7.5 to 15 cm) thick on top to weigh down the lightweight cardboard so it doesn't blow away. Soak the whole area with water. As all of these materials decompose and settle onto the earth below it, they feed the soil.

Undertaking new plantings requires thought before action. Removing existing vegetation prior to planting helps prevent weed issues down the line. Densely planting is another way to prevent weeds. This new meadow planting will leave little room for weeds as the plants mature.

Avoid mechanical tilling whenever possible as it disrupts soil life, brings up buried weed seeds, and releases carbon sequestered in the soil. However, one-time tilling may be helpful when creating a new bed that is too large to establish by other means.

You can plant into this right away, but letting it sit for four to six months gives you a better base to work with and allows the soil web to establish itself. Use this method any time of year.

SOLARIZATION

Another means of eliminating the grass, solarization uses a clear plastic sheet to essentially "bake" the ground underneath. Starting with closely mowed grass and wet ground to create a greenhouse effect, cover the area with thin plastic sheeting, and bury the edges so it's completely sealed.

Solarization works best during the hottest parts of summer, harnessing the most direct sun. In two months, the vegetation underneath will be killed, and you can start your garden.

Downsides to this lawn-conversion method are that you're using plastic, which isn't ecologically minded, and the solarization kills or disturbs the soil life in the upper portion of the soil. It'll take a little while for your soil life to rebound.

TILLING

Using gas-powered equipment isn't as ecological as the human-powered solutions above, but it is efficient. In chapter 3, you read about how to foster life in the garden. Minimizing soil disturbance is one of the tenets, and tilling upends that idea. A thorough tilling to do away with persistent turfgrass will damage soil life, yes, but if this is just a one-time tilling, the soil web can rebuild and be as strong as ever in just a few seasons.

Another downside to tilling is that it stirs up the weed seeds held deeper in the soil. As the seeds are exposed to light, they can germinate. Work fast in covering tilled ground with mulch or another means of blocking the light, such as cardboard or plywood, until you're ready to work in the garden.

Tilling might make the most sense if you're converting a large garden space. It becomes harder to manage sheet mulching or plastic-sheet solarization as the coverage area gets larger. Be discerning in your timing. If the ground is too wet, tilling creates large chunks of soil that are nearly impossible to work with. If it's too dry, you'll have a dusty mess and risk erosion.

A tiller is a small gas-powered piece of equipment, about the size of a push mower. If you don't have one in your garden shed, you can rent one from an equipment-rental company. Your Extension Service office or local tool-lending library might also have one for you to borrow.

LEAF CLEANUP

FOR MUCH OF NORTH AMERICA, one of the biggest lawn and garden events is when deciduous trees drop their leaves in the fall and out comes an army of people with rakes and leaf blowers in hand to gather them into piles, pack them into bags, and haul them off to either the garbage or—hopefully—municipal composting facilities. There has been a movement against such a big leaf cleanup process that's known as "leave the leaves." It advocates for letting the fall leaves remain where the wind drops them. Sometimes this will work out just fine, and other times, not so much. We'll go through some situations where you will want to clean up leaves. When you do, you'll want to keep those leaves on your property. Fall leaves are full of nutrients to fuel plant growth, and many insects hibernate tucked down in the protective cover of leaves. If you can relocate them rather than throwing them away, it is the most ecological solution.

DRIVEWAYS AND WALKWAYS

One place where you should always clean up fall leaves is from walkways and driveways. From a human point of view, wet, decomposing leaves on hard surfaces can become slippery, so it is a safety issue. From an ecological point of view, fall leaves are full of nutrients that will be released as they break down. When stacked on the soil, those nutrients go into the ground to fuel plant growth. When piled on a sidewalk or driveway, those nutrients head down storm drains and into waterways where they can fuel algae blooms and harm wetland ecosystems.

LAWNS

Fallen leaves can smother and harm lawn grasses, but only if the leaf layer is too thick. A good rule of thumb is that if the leaves cover more than one third of the surface of your lawn, they are too thick and will need to be managed to provide the grasses with enough light to survive. Very large, thick leaves, like sycamores and some oaks, can have a bigger impact than small, delicate leaves, such as those of a honey locust. When leaves are too thick on a lawn, you can either rake them aside to somewhere where they won't smother anything, or run them over with a lawn mower to shred them and allow the grass to grow up through them.

One of the worries expressed by many gardeners about not removing leaves from garden beds is that their plants won't be able to push up through the leaves. But, if small woodland ephemerals can do it, most cultivated garden plants will have no trouble pushing up through a leaf layer left in the garden.

Leaves left in place in garden beds provide insulation through the winter and form a natural mulch layer that is useful for reducing weeds and watering needs. Leaves can be pulled away from the crowns of smaller plants and evergreens if necessary.

PERENNIAL BEDS

When fall leaves accumulate in our perennial beds, it can be hard to know what to do. In most cases, a layer of leaves is beneficial, adding insulation through the winter, releasing nutrients as they break down, and forming a natural mulch layer. However, very thick leaf layers, especially in corners where the wind collects them, can smother some perennials and kill them. So how do you know if you can leave the leaves in place? There are a few rules of thumb. If evergreen plants are completely covered, they will probably die, so gently brush away the leaves from their crowns to allow these plants to photosynthesize over the winter. The smallest of plants, those less than 6 inches (15 cm) tall and those that compose your ground cover layer, are also the most likely to be smothered by thick leaves. Gently raking leaves off those low-growing plants and into parts of the garden with taller perennials will ensure these little beauties thrive.

Most larger deciduous perennials will push through a thick layer of leaves easily, but some may struggle. One easy way to figure out just how much a perennial can handle is simply to observe them in spring. As the weather warms up and plants come into growth, walk through the garden. If you see green shoots pushing up through the leaves, then everything is fine. If nothing is showing above the leaf layer, gently pull the leaves away. If, under them, you see somewhat twisted, white or pale green shoots, that's a clear sign the leaf layer is too thick. Pull the leaves away now so the plants can emerge, and make a note of which perennials needed your help so that next fall you can keep the leaves off their crowns altogether.

AESTHETICS

Finally, you may opt to clear away some leaves simply because you like the look of a tidy garden, especially if it is just in small areas. In natural woodlands, deep piles of leaves accumulate in some areas, and in other spots the wind clears them away, leaving bare soil, and different plants and insects adapted to thrive in both conditions. Mimicking this variation in your landscape by clearing some areas and piling leaves deeper in others can allow you to create a tidier, more manicured look where you want it, while also creating a diversity of conditions for different species.

PRUNING

DEPENDING ON YOUR GARDEN and situation, you may almost never get out the pruning shears to trim your plants. On the other hand—especially if you have a very small garden—pruning may be a central part of your maintenance practices. Pruning is a useful tool to both maximize the health of our plants and help plants and humans live together safely and comfortably. Often smart pruning can be the difference between a plant being a long-lived and vital part of your landscape and having it cut down and removed because it was growing into walkways or up through electric power lines.

It's always best to hire a licensed arborist to prune tall trees or those near houses or power lines.

TREE HEALTH

The most important pruning you will do is to maintain the health and longevity of trees in your landscape. While many trees will never require pruning, there are a few key things to look out for and correct, especially as young trees are growing.

A healthy tree should have a single, straight trunk reaching up and smaller branches coming off of it as it develops. Having a single trunk is important to the structural integrity of a tree. If the tree splits into two or more trunks, the point where the two trunks come together is a structural weak point that is likely to eventually split in a storm or heavy snowfall. A key to making sure a tree develops a single strong trunk is pruning it to a single leader when it is young. A leader is the central, tallest branch on a young tree, and it is what will eventually thicken and develop into the trunk as the tree matures. If two or more branches are equally tall, they'll both develop into trunks, setting up the tree for failure in the future. Trees will naturally grow with a single leader, but if the tip of that leader gets damaged or nibbled back by a deer, it will often cause multiple leaders to develop. Unfortunately, some nurseries will prune back the central leader of a young sapling on purpose to make it branch out earlier so it looks fuller in the nursery, but that can cause problems down the line. So when shopping for trees, look for narrow saplings with a single leader and not bushy ones with lots of branches. If a young tree in your landscape has multiple leaders, prune all but one of them off so the remaining one can establish as the main trunk.

If you have a mature tree with multiple trunks, consult with a licensed arborist about the best course of action. It may be possible to prune the tree to reduce the chance of the trunk splitting. It may be in a location where you can leave it to live as long as it can; if it is somewhere where it might fall and harm a structure or people, you may need to have it taken down.

Though trees are fundamental to healthy ecosystems, key to moderating temperature, and absolute powerhouses when it comes to feeding local ecosystems, they can also be dangerous when in the wrong place. Trees close to houses and hanging over power lines need to be carefully monitored and cared for to ensure they don't become a safety hazard. The most basic pruning in this case is to cut off dead or dying limbs before they can fall and hurt something or someone, and you may consider hiring a professional to prune away healthy branches that hang over power lines or seating areas. If possible, move any pruned dead wood to an out-of-the-way place on your property to serve as a home for wildlife.

SIZE CONTROL

Maybe you planted a little shrub without looking at the tag, or perhaps you inherited a shrub that someone else planted. Whatever the reason, many gardens have shrubs and small trees in places where there is no room for them to reach their mature size. Careful pruning can help keep a shrub in check and prevent you from needing to remove it altogether. The key to pruning for size control is to start early and prune a little bit frequently. Quickly and lightly pruning back the tips of branches beginning to grow out of their proper space will be an almost invisible way to get the job done, and the plant won't be harmed by the pruning. If you wait until branches have grown halfway across the sidewalk, cutting them all of at once will leave an ugly open space on the plant and can harm the plant long term. If you do need to do a major size reduction on a shrub that is vastly overgrown, it is best to do it in stages, pruning back a little at a time every few years, and giving the shrub time to recover before removing more wood.

SHEARING

Shearing shrubs is a very traditional method of pruning. It involves neatly cutting off all of the new growth to form the plant into a geometric shape like a ball or cube. Though a tightly sheared hedge looks very artificial, there is no ecological reason not to do it. A sheared hedge is still good nesting habitat and provides just as much food for caterpillars. In fact, if you are struggling to get your ecologically sensitive landscape accepted by uptight neighbors or an HOA, keeping some shrubs tightly and neatly sheared can be a way to make your landscape look more intentional and cultivated without losing any of its ecological value.

The one golden rule of shearing shrubs is to make sure you always make the base of the shrub wider than the top—even if only by a little bit. If the top is wider than the bottom, it will shade out the bottom of the shrub, causing the lower branches to die back. These light-starved lower branches result in a rangy, open-looking shrub instead of a tidy, dense block of foliage. This is particularly critical on the north side of a shrub or hedge, which will be more shaded.

Shearing is a pruning method that involves cutting off all new growth to form the shrub into a specific shape. If "neat and tidy" is required by your HOA, or if it's your personal preference, many native shrubs, like this ninebark, can be pruned by shearing.

ARBORIZING

Another pruning method to manage an overgrown shrub is called *arborizing.* This means to prune off most of the lower branches, leaving just one or a few as trunks reaching up to a canopy of branches above. This can be a way to open up planting space under a tall shrub and is a great solution for tall shrubs that are blocking a sight line or window. Instead of trying to prune it down to be short again, pruning away the lower branches can open up the view and transform the shrub into a small tree.

PRUNING PERENNIALS

Pruning isn't just for woody trees and shrubs. There are some useful pruning techniques you can use on your herbaceous perennials as well. One of the most useful is an early summer cutback, also called the *Chelsea Chop* after the famous British garden show. You may also hear gardeners call this process *pinching.* This is a great way to reduce the height of a plant and the need for staking tall summer and fall-blooming perennials. In late spring, take a pair of shears and lop off the top third of a perennial. This will cause it to branch and grow shorter and denser, making it less likely to flop over or require staking. Summer pruning can be a really great way to tame sprawling native perennials that can grow too tall and lanky in a fertile garden setting and give them a more tidy, cultivated look without sacrificing any of their beneficial contributions to your garden's ecosystem.

Pinching is a way to prune perennial plants that increases the number of blooms and keeps the plant more compact. Stems are pinched or cut back to a place just above where leaves emerge. In this photo of a recently pinched bee balm stem, you can see two new shoots emerging from where there was once just one shoot, effectively doubling the number of blooms that will be produced.

PREVENTING SEEDING

Whether it is a perennial, tree, or shrub, you sometimes may want to prune to prevent seed production. Usually seeds are a good thing as they are a valuable food source for many birds and insects, but not always. Sometimes a beautiful plant can become a bit of a weed by seeding around too aggressively in your garden, especially if it is an invasive species. Ideally, you will not have any invasive species in your landscape to begin with, but you may have inherited some, and you can't always remove them right away. As you work your way through carefully replacing invasive species with something better, you can stop them from multiplying by preventing them from setting seed.

The simplest way to do this is by deadheading—cutting off the flowers right after they fade. For some trees and shrubs, you can even do more extreme pruning. Many shrubs will, if cut down right to the ground, shoot up again with new growth. By doing this annually, you can keep them from flowering and setting seed, while keeping their ground-stabilizing roots and habitat-giving branches in the garden until another plant takes their place.

If you choose to grow nonnative plants that are prone to reseeding, it's essential that you cut off the spent flowers prior to the seeds developing to prevent their spread. Rose of Sharon is a popular nonnative shrub that will self-sow everywhere if the plant is not deadheaded.

CUTTING BACK

In addition to deadheading and pruning early in the year, the other big traditional task for maintaining perennials is to cut back and remove their dead stems either in the fall after they go dormant or in the spring before they push out new growth. The practice of cutting back and doing a big fall cleanup is a purely aesthetic exercise, done to make the garden look tidier and more cultivated. Perennials don't need to be cut back to be healthy, and leaving old stems in place can be beneficial to plants as they provide a little insulation during winter cold.

Leaving old stems and leaves in place is also great for the other things living in your landscape as many insects hibernate in the winter in the leaf litter at the base of plants or in the hollow stems of plants (see page 93). These stems also act as homes to certain insects when rearing their young. So, the best practice from an ecological point of view is to just leave everything as it is, and let nature take its course, but that may not be practical in every garden. Here are a few strategies you can use to keep the garden looking tidy while still leaving old stems standing as habitat.

First, you can cut back plants only partially. Rather than cutting plants down to the ground, you can leave half to a third of the stem length in place, which provides plenty of habitat, but also gives a tidier appearance. Plus, the old stems will be more quickly covered by new growth in the spring.

You can also practice a thorough cutback in parts of the garden while letting other areas be wilder. Perennial beds in the front of the house, along sidewalks, and in other high-visibility areas may be good places to practice more traditional maintenance and cut back dead stems completely, while more out-of-the-way areas can be left to develop without any trimming or cutting back. Compromises like this can be a way to grow an ecologically sensitive garden without making neighbors or HOAs angry.

EDGING

Taking a sharp spade or edger and carefully cutting a perfectly neat, tidy edge between a perennial bed and the lawn can be a powerful tool in making an ecologically sensitive garden look tidy, cultivated, and acceptable to the general public. A crisp edge communicates without compromising ecological value that the planting is intentional and cared for (remember the term *cue to care* from earlier?). If going out regularly and edging a bed with a spade is too much work, you can get the same effect with a metal or stone edge, creating a formal frame for the garden it surrounds.

Cutting back perennial plantings is purely an aesthetic practice; if you don't cut back your plants, it will not harm them. However, if you do opt to cut perennials back, consider doing so as late in the spring as possible to provide fall, winter, and early spring habitat for insects. Leave several inches of stem stubble standing after cutting for habitat throughout the growing season.

WEED MANAGEMENT

MANY AN AMBITIOUS gardening scheme has been brought to its knees by weeds. Dreaming up a new bed, picking out plants, and planting them—that is the fun part. What comes after is the sometimes-tedious process of keeping weeds from taking advantage of all that open space and carefully prepared soil to overwhelm your garden. And once weeds get growing, they get worse by the day, easily spinning out of control and making many a new gardener give up soon after they start. Luckily, weeds can be very manageable with a little knowledge and planning.

There are many invasive weeds capable of crowding out desired plants, including the chickweed and garlic mustard shown here. The upside: Both of these weed species are edible.

WHAT IS A WEED?

The definition of a weed is pretty simple: It's a plant where you don't want it. That means that the very same plant can be a weed in one context and not a weed in a different one. Creeping bentgrass is a weed when it lands in a vegetable garden, but it's a very desirable part of a traditional lawn. Generally, a plant has a high chance of being considered a weed if it grows very rapidly, if it has thick, dense growth that can shade out and kill other plants, if it spreads quickly either by runners or seed, or if it is has sharp thorns or other unpleasant characteristics. Poison ivy, for example, would be classified by most people as a weed despite being a native vine with beautiful fall color. The awful itching it causes trumps its many virtues.

But sometimes rethinking your definition of a weed can be helpful. Just because you didn't intentionally plant something doesn't mean it is a weed. Some things we are tempted to think of as weeds can be left alone to become another part of your landscape and ecosystem. This applies especially to low-growing species that don't shade out other plants.

WHERE DO WEEDS COME FROM?

We go to great lengths to acquire, propagate, and care for the plants that we want in our gardens and landscapes, but the weeds just seem to show up, as if by magic. Understanding where weeds come from is key to successfully managing them with the minimum of effort and disruption.

The vast majority of weeds in most gardens—especially newly planted gardens—comes from what is called the *soil seed bank*. The seeds of many species of plants, especially the fast-growing ones that are often classified as weeds, can live for decades in the soil, waiting for just the right time to germinate. Some species can even stay viable in the soil for over 100 years. When conditions are right, these seeds germinate and leap into growth. Many of these extremely long-lived, weedy seeds have evolved to take advantage of disturbed environments. When a tree blows over or a flood washes through, leaving bare, open ground, they spring into life and quickly grow to fill the space. That means those same seeds are waiting to germinate when you clear away lawn grass to make a new garden space. Nature abhors a vacuum, and there are many plants that have evolved to quickly take advantage of open ground and open areas.

In addition to the weeds that germinate from seeds in the soil seed bank, new gardens sometimes inherit perennial weeds that were there before and were not completely killed in the process of making a new garden. Many plants can regenerate from even quite small root fragments left in the soil. The good thing about most of these kinds of weeds is that they are slower growing from seed, and so once eliminated, they are pretty easy to keep at bay by regular weeding to keep new seedlings from getting established. Avoiding perennial weeds is really a matter of ensuring the prep of a new garden space was thorough, and being aggressive in removing any weeds that pop up. While fast-growing weeds from seed can be removed by pulling them out with your fingers or even sometimes cutting them off at the soil level, perennial weeds are best dealt with by getting a shovel and carefully digging out the whole root system.

New weed seeds get carried into gardens all the time as seeds blow on the wind or perhaps when a bird stops to relieve itself after a big berry dinner. The reality is, however, that relatively few seeds move around this way. Most plants distribute seeds only a few feet away from the parent plant, so while a few species of high-flying weeds may float into your garden on a breeze—like the infamous dandelion—the biggest source of new weed seeds in your garden will be the plants in your garden itself or in the immediate area around it.

Perennial, deep-rooted weeds, like this Canada thistle, are particularly troublesome to remove in an ecologically sensitive manner.

Sometimes, in a mature garden, the worst weeds are things that we ourselves planted there when the garden was younger. Novice gardeners are usually impatient and eagerly snap up plants that are advertised to grow rapidly and spread. Those rapid growth habits that are so helpful in a new garden can quickly become a problem as a garden matures and the rapid spreaders start overtaking their slower growing, but maybe more desirable, neighbors. There are a few strategies to manage these types of plants. First, of course, is not to plant them in the first place, instead choosing compact, nonspreading forms for your garden design. The other option is to place plants with similar growth habits together. Two or three fast-spreading species grown together will quickly fill a bed and be vigorous enough to stand up to their neighbor's aggressive habits. Or, you can closely monitor the more aggressive species and once or twice a year cut back or dig out shoots that have popped up where they are not welcome.
So, how do we then keep all these sources of weeds from becoming a problem? There are a few different methods that can really stop weeds in their tracks, giving you more time to enjoy your garden and decreasing the time spent pulling up plants you don't want.

DENSE PLANTING

One way to limit weed problems is simply to not give them much room to grow. Planting densely so the plants you want cover the ground completely will ensure that any weed seeds that did manage to germinate will be shaded out and unable to mature. Dense plantings will also quickly produce a natural mulch layer to further suppress any weed growth.

You can also think about planting density in time as well as space. If your garden beds are packed full of plants in the summer, but bare in the winter when perennials are dormant, you might consider adding a low, evergreen ground cover layer or a winter annual cover crop to ensure the soil is always covered and weeds never have a chance, any time of the year.

Plant densely! Leave no room for weeds.

BRING IN CLEAN SOIL

One way to sidestep the issue of the soil seed bank is to bring in clean, weed-free soil to your garden. This isn't practical for large areas, but can be a good solution for small, intensively cultivated spaces like a vegetable garden. Building a raised bed and purchasing a soil mix that is weed free can mean you'll only have to deal with the few weed seeds that blow in, making it a much easier space to care for. This is particularly helpful in vegetable gardens where frequent harvesting and replanting makes it harder to keep the soil covered all the time. Be sure to amend this soil with finished compost to kick-start the introduction of soil biodiversity.

AVOID SOIL DISTURBANCE

Given that soil disturbance triggers the growth of seeds in the soil seed bank, minimizing such disturbance can reduce weed problems. Ways to do this can include purchasing plants grown in smaller pots so you can get them in the ground without having to dig a huge hole, cutting down rather than pulling up weeds when possible, and managing water flow to prevent erosion.

KEEP A REGULAR SCHEDULE

One of the absolute best tools in weed management is setting aside regular time in your schedule to walk through your landscape and pull weeds. Weeds are kind of like compounding interest, but in a bad way. They get exponentially worse every day you leave them unattended. You can pull out a small, newly germinated weed in two seconds without disturbing the soil or other plants. Wait a week, and it will be a big weed that requires a tool and will disturb the soil when you pull it, awakening other dormant seeds in the soil seed bank. Let it go even longer, and it will start shading out and weakening the good plants in your garden, meaning they cover the soil less effectively, creating even more openings for weeds to grow. And let a weed flower and go to seed and you will have countless new weed seeds added to the soil seed bank that you will have to pull in the future.

If you make a habit of walking through the garden every week and pulling the few weeds you find, you'll have a relaxing weekly stroll through the garden and a virtually weed-free space. Put it off for a few weeks, and you'll suddenly have a big, difficult, unpleasant chore on your hands. Weed often—the more frequently, the better.

A LITTLE AT A TIME

The final tip to ecological weed management is to recognize that new garden plantings will be a lot more work to keep weed-free than established ones. New plantings have a lot of exposed, disturbed soil and a seed bank full of weed seeds ready to spring into life. As a garden matures, plants fill in, natural mulch builds up, and the soil seed bank gets depleted by regular weeding, weed problems can drop to almost nothing. So, if you want to put in a huge garden, don't do it all at once. Start by creating and planting just one small section and pay attention to how much time it takes you to keep it cared for and weed-free. If you are feeling relaxed and on top of it, you can add another small area. Once that feels like enough to do, give the garden some time to mature and the weed presence to drop before starting another area. Doing a little bit at a time will ensure each new garden section gets all the care it needs during establishment, and weeding remains a quick and simple task rather than an overwhelming disaster.

When planting, try to disturb the soil as little as possible so weed seeds are not brought up to the soil surface.

BUILD *Better* SOIL

Meadow plants, like this purple prairie clover (*Dalea purpurea*), thrive in leaner, less nutrient-rich soils with a lower organic matter content.

Woodland plants, like this goldenseal (*Hydrastis canadensis*) thrive in richer soils with high organic matter content.

HEALTHY SOIL IS A HALLMARK of ecological gardening. The health of your soil, in fact, dictates the health of your garden.

As we've already briefly discussed, within soil is a living, breathing community of organisms—the soil's own web of life. Everything you do above ground has an impact on the organisms living below ground. In nature, processes already exist to support soil life and create better growing conditions for plants. It's ironic that some garden practices actually inhibit these natural workings. It's up to you to support this natural soil-building process.

While some native plants thrive in leaner, less nutrient-rich soils (many prairie and desert plants, for example), other native plant species prefer soils higher in organic matter (woodland plants, for example). It's important to tailor your soil-building efforts to the plants you are growing and the climate in which you live.

UNDERSTANDING THE NUTRIENT CYCLE

Nature's means of creating healthy soil is called the *nutrient cycle,* and it's every gardener's best friend. The simplest example of the nutrient cycle is in the leaves of a deciduous tree. In the winter, the tree is dormant, and its branches are bare. The tree's energy is stored in its roots, which exchange water and nutrients underground. As the daylight increases and weather warms in the spring, the tree's leaves emerge. Through summer, the tree's above-ground activity is on full display, with branches and leaves growing and maturing. Come fall, the length of daylight is noticeably shorter, the weather cools, and the leaves change color and fall from the tree. It's now winter, and the cycle begins again.

At no point in this tree's seasonal sequence did someone add a bag of fertilizer or a wheelbarrow of compost, but the tree's growth tells us it's still accessing nutrients. This can go on for hundreds of years. The tree is fed by the activity of the organisms in the soil, which are fed by the fallen leaves as well as by the nutrients that visiting insects, birds, and wildlife leave behind. These items also feed all of the woodland plants growing beneath the tree. This is the nutrient cycle.

In a forest ecosystem, the nutrient cycle involves the decomposition of fallen leaves each autumn, as well as other decaying plant and animal materials.

Nutrient Cycle Phases

The nutrient cycle has four phases:

- Plants take up nutrients from the soil via their roots.
- Plants use the nutrients to grow their stems, leaves, flowers, fruits, and seeds. These components all store nutrients.
- Animals eat the plant parts, extract the energy and nutrients they need, and deposit the rest on the ground in their waste. Plant materials fall to the ground in the annual cycle of life and death and return their nutrients back to the soil. When animals die, their decaying bodies also nourish the soil.
- The insects, fungi, bacteria, worms, and other forms of soil life break down the animal and plant matter into components that are useable by plant roots.

All parts of this never-ending cycle happen concurrently. The nutrient cycle is here for the ecological gardener to harness.

Harnessing the Nutrient Cycle

The process of nutrient cycling is evident throughout smart gardening practices, which you'll read about in this chapter. Still, when humans become involved in natural processes like this, there's room for error.

There are a few ways you can boost the nutrient cycle for plants that require increased soil fertility:

- Use mulch made from organic materials. This breaks down over time and returns its nutrients to the soil.
- Compost. Whether you leave your dead plant material in the garden to compost in place or you add it to your compost pile, it will decompose to feed the soil.
- Carefully select soil amendments. The right amendments at the right time can support the nutrient cycle (see page 184).

There are also a few ways you can interrupt the nutrient cycle:

- Covering soil with plastic-based sheet mulch, such as landscape fabric or plastic weed barrier, slows or eliminates the nutrient transfer from above ground to below. It may also heat the soil too much and destroy the soil life close to the surface.
- Removing organic matter from the garden, whether by harvest or pruning, and not returning it in the form of compost does away with a valuable, yet free, nutrient resource.
- Any practice that damages soil life, such as using pesticides or continually disturbing the soil, disrupts decomposition work.

When growing vegetables, adding organic matter such as compost to the soil improves soil structure and adds nutrients to support the growing plants.

Soil organic matter. The point of the nutrient cycle is to create and support soil organic matter: the living or once-living components in the soil. You may have already realized, or you will soon realize, that soil organic matter fills a lot of gaps in ecological gardening. Whether you have soil that's too dense and impermeable or soil that's too loose with little water retention, increasing the soil organic matter can fix that. When you have soil with little invertebrate and microbial activity, soil organic matter can give it a boost. Soil with little mineral content also benefits from soil organic matter, which can aid in soil mineral decomposition. Your goal should be to maintain a thriving soil web full of soil organic matter.

As you work on improving the soil in your garden, there are forces that both increase and decrease the percentage of soil organic matter. Adding the proper soil amendments and keeping soil covered with plants and organic mulches put marks in the plus column. In the negative column are erosion or other soil loss, harvesting produce or removing plants, using chemical fertilizers and pesticides, and tilling. This is part of the ecosystem's give and take.

For vegetable gardens and traditional flower beds, the recommended minimum soil organic matter is 3.5 to 4 percent. That means that, by weight, organic matter makes up 3.5 to 4 percent of your garden soil. For meadow plantings and other garden areas largely comprised of native plants, a lower percentage of soil organic matter is optimum because many of those plants perform better in leaner soils.

GET TO KNOW YOUR SOIL

In building soil, you need to know what you're working with to start. The manufacturers of soil amendments like "plant food" and fertilizers want you to believe you need products like theirs to have a healthy garden. Some of these products do help under the right circumstances. Before you can make decisions about amendments and fertilizers, ecological gardening asks you to begin with what you have. You may recall the discussion on soil structure in chapter 4, where it was noted that good soil structure is responsible for the loose, crumbly soil gardeners covet. It is also filled with organic matter and soil life that helps hold its aggregates together, allowing for the easy passage of roots, water, air, and nutrients.

The best soils allow for a balance of water drainage and retention: enough structure to provide stability to plants and their root systems; porous enough for oxygen to pass through; and can hold nutrients. These soils are described as *friable*. While some types of soils are more desirable than others, there are positives and negatives to all soil types. Whatever your soil type, know that this is what you have been given to work with, and you can improve it over time with the cultivation of soil organic matter.

Get a soil test. To determine the nutrients your soil needs to best support your plants, order a soil test before you order any growth agent that comes in a bottle or bag. A soil test measures the nutrients, micronutrients, soil organic matter, and pH of a patch of ground. This is important because you don't want to waste time, money, or resources on amending the soil with substances it doesn't need. Too much of a nutrient is just as detrimental to your plants and soil life as not enough of the same. Likewise, a pH that's too acidic or too basic can prevent plant roots from accessing certain nutrients, regardless of how rich the soil may be. Ecological gardening is all about balance—this goes for balanced nutrients and micronutrients as well.

If you're in an urban area or one that's had a history of industrial use and you plan to grow fruits, vegetables, or herbs, be sure your soil test looks for heavy metals and contaminants.

You have a few soil test options:

- Purchase a simple kit from garden-supply retailers. These might not measure the full range of nutrients and micronutrients, but you'll learn your results immediately.
- Ask your county Extension Service office, which might offer soil tests for only a few dollars. These tests can take a few weeks or longer, because they usually need to be sent to a state lab.
- Send away your soil samples to a commercial testing lab. This is the most expensive but most thorough option, and you usually receive your results within a week. For a little extra money, an agronomist at the lab can analyze the soil test results and make recommendations to correct any imbalances.

SOIL AMENDMENTS VS. FERTILIZERS

The terms *soil amendment* and *fertilizer* are often used interchangeably, though there's an important difference. Now that you understand the nutrient cycle, you may more clearly understand the distinction.

A soil amendment is applied directly to the soil with the intention of improving the soil health or structure and adding organic matter. A fertilizer may be applied to the soil or directly to the plant and is meant to provide plants with a nutrient boost with little regard for the soil itself. You might think of fertilizers as plant food, whereas soil amendments are soil food.

Healthy soil creates healthy plants. In ecological gardening, you're striving for long-term resilience rather than the quick fix that fertilizers provide. As you build your soil through the use of soil amendments and other techniques discussed in this chapter, you will have less need for fertilizer, if any at all.

About fertilizers. While fertilizers are not necessary—and may even prove detrimental—for native plants that are matched to their proper ecosystem, if you are an ecological gardener who grows a vegetable garden or ornamentals in containers, the following information may prove useful. Common fertilizers supply the three main nutrients that plants require: nitrogen, phosphorus, and potassium. Some also provide secondary nutrients and micronutrients, which plants need in smaller quantities. Secondary nutrients are calcium, magnesium, and sulfur. Micronutrients include boron, copper, iron, and zinc.

On a fertilizer label, three numbers represent the main nutrient content. An "all-purpose" fertilizer labeled 10-6-4 has 10 percent nitrogen, 6 percent phosphorus, and 4 percent potassium. This is a complete fertilizer, but if your soil already has enough of any of these nutrients to support your plants, you're wasting your money and potentially polluting the soil and groundwater by adding more.

There are two main types of fertilizer products: synthetic fertilizers, which are made in a laboratory and should not be a part of ecological gardening practices, and natural fertilizers, which are derived from various natural sources, including plants, animals, and minerals.

Examples of common natural fertilizers include:

- Alfalfa meal, high in nitrogen
- Bone meal, high in phosphate and calcium
- Kelp, high in potassium and a source of trace minerals
- Manure, providing a low level of all main nutrients
- Various branded products that provide nutrients, secondary nutrients, and micronutrients

Proper use of these and other fertilizers underscores the need for gardeners to understand plants' growth habits and nutrient needs.

In addition to being characterized by the nutrients they supply, fertilizer products are also classified by their mode of action:

- Quick-release fertilizers give your plants the boost that your afternoon coffee gives your productivity. It's an immediate, short-lived nutrient shot. This is sometimes necessary, such as for a nutrient deficiency that's causing a plant disease. Because plants' needs change depending on their stage of development, it's important to use quick-release fertilizers at just the right time in the right concentration. Nutrients not taken up by the plant can leach out of the soil, especially in the case of wet weather or overwatering. Quick-release fertilizers may be in a liquid or granular form. These are generally the least expensive of the fertilizer types, but in many cases, they are synthetic products that don't belong in an ecologically sound garden.

- Controlled-release fertilizers may contain the same nutrients as quick-release fertilizers, but they're in a form that the plants can't use all at once. The nutrients are coated to release in the amount and at the timing needed by the plant's growth cycle, dependent on temperature and moisture conditions in the garden. Most often, an industrial, synthetic process is used to create the controlled-release mechanism, even if the nutrients themselves come from an organic source, again making them an unwise choice for your ecological garden.

Natural fertilizers, such as this kelp meal, provide nutrients to plants as they are broken down by soil microbes.

Careful and judicious applications of natural fertilizers can be helpful for giving plants a boost of nutrients if a soil test shows they are necessary.

- Slow-release fertilizers are those made available over time by the microbial decomposition process. Some of these are manufactured and should be avoided in the practice of ecological gardening, but there are also natural sources of fertilizer that are considered slow release. Manure is a natural slow-release fertilizer, with 30 to 50 percent of its nitrogen released as it's broken down in the first year, and decreasing amounts released after that.

Slow-release and controlled-release fertilizers sound similar, but the difference is that controlled-release fertilizer nutrients are manufactured to be made available to plants over time when specific environmental conditions are reached; slow-release fertilizers are dependent on microbial activity in the soil, which is harder to control but more eco-friendly.

When to use fertilizers. Gardeners interested in growing the showiest begonias and winning the largest-pumpkin contest at the state fair use fertilizers to pump up their plants' production potential. Fast-growing, big-producing plants require a lot of nutrients, which fertilizers can supply. Even regular vegetable and fruit production draws nutrients from the soil that will need to be replaced. Vegetable gardens may have more of a need for regular fertilizer applications because you need to replace what you remove. Annual plants growing in container gardens can't take advantage of the nutrient cycle, being removed from the earth, and may also need regular fertilizers to maintain plant growth and health.

However, the slow and steady growth that most ecologically minded native-plant gardeners look for has no need for fertilization. Proper soil conditions provide all the nutrients these thoughtfully selected plants need.

In landscapes combining edibles and ornamentals, fertilizers can be applied specifically to the edible crops without the need to apply them to the ornamental species.

Soil tests and plant-health observation are two ways you'll come to know of your garden's fertilizer needs:

- A soil test is the more straightforward of these indicators. Find a natural fertilizer product that supplements the lacking nutrients and apply it according to label instructions.
- Plant health is a more challenging indicator, as your plants don't have nutrient gauges on their leaves. Instead, you have to read the signals the plant offers, such as yellowing leaves, stunted growth, lack of fruit production, and other anomalies. You may need the help of a plant-health guide, a trusted website with science-based advice, or your Extension Service agent to determine the problem and solution.

From an ecological gardening perspective, healthy soil creates healthy plants. You're striving for long-term resilience rather than the quick fix that fertilizers provide. As you build your soil through the use of soil amendments and other techniques discussed in this chapter, you should end up in a place where fertilizer is completely unnecessary.

About Soil Amendments

Because soil amendments are meant to build soil, they have less to do with nutrients and more to do with physical characteristics.

Soil amendments include:

- Coconut coir, a byproduct of coconut processing, to improve water permeability
- Compost to build soil structure, lighten heavy soils, and feed the soil web
- Lime, a mineral, to increase soil pH (making it more alkaline)
- Manure to increase soil organic matter, feed soil life, and add nutrients
- Worm castings, or vermicompost, as a source of beneficial soil microorganisms, enzymes, trace elements, and nutrients
- Biochar to increase soil pH and improve water and nutrient retention

Compost is the ideal soil amendment for building healthy soil.

As with fertilizers, these soil amendments are naturally occurring, but they could be mixed with or processed using synthetic materials or treatments, so it's important to know the source.

When to Use Soil Amendments

With the goal of building healthy soil, you'll reach for soil amendments more often than you will fertilizers. Many soil amendments aim to increase soil organic matter, which your garden may or may not need, depending on the plants you are growing.

While soil amendments create all of this good, some native plants simply prefer soils with lower nutrient content, and for those plants that do require more organic-matter-rich soil, applying too many amendments too fast can backfire. You can end up with high salt content in your soil or too much nitrogen or another nutrient. Over-amending your soil can also lead to too little nitrogen available to your plants, which is counterintuitive, but there's good reason. Microorganisms require nitrogen to break down high-carbon soil additives, and they'll temporarily pull the nitrogen from the soil, away from your plants, to do this work.

Soil amendments might be a regular part of your ecological garden routine, particularly if you are growing a vegetable garden (less so if your focus is on native perennials and shrubs). Follow application instructions to build your soil over time, in the way that's best for the plants growing there. As you improve soil structure and bring the soil web to life, it's likely you'll eventually have no need for soil amendments.

Choosing Fertilizers and Amendments

While the fertilizers and amendments listed as examples in this book are naturally occurring, they could be mixed with or processed using synthetic materials or treatments. Be sure you understand the labels, and do your research into products and brands before making your choice.

One way to know you're applying fertilizers and soil amendments without synthetic additives or processing is to look for the Organic Materials Review Institute (OMRI) label. OMRI is a nonprofit that independently reviews agricultural products, measuring them against the USDA National Organic Program standards. It makes no difference that your garden isn't certified USDA Organic; this label tells you these products could be used in a certified-organic setting and therefore don't contain prohibited synthetic substances. This approval isn't the final measure of an ecologically safe product. There may be OMRI-listed products that don't match your ecological-gardening principles, and there are products out there that haven't been OMRI reviewed that would make the cut.

FOREVER CHEMICALS

Here's something our grandparents didn't think about when they started their gardens: Per- and poly-fluoroalkyl substances (PFAS), also known as *forever chemicals*. PFAS were invented in the 1930s. Their industrial use began with the atomic bomb in the 1940s, and from there, they've been an integral component of nonstick cookware coatings, water-repellent fabrics, firefighting foams, household cleaning products, and more.

PFAS take thousands of years to degrade, meaning we're stuck with them. This is bad news because they have been shown to cause negative health effects in humans and animals, ranging from increased risk of some cancers to increased cholesterol levels and decreased fertility.

Some states are passing legislation to limit the production and use of PFAS, and yet, PFAS are everywhere. They leach into the environment from industrial sources, landfills, and water-treatment facilities, so limiting your exposure is challenging. But you can do your best to keep PFAS out of your garden.

One major source of PFAS is biosolids, which are byproducts of wastewater treatment plants used in commercial fertilizers. Biosolids are prohibited by USDA NOP standards for crop production, so any compost or manure-based product that's OMRI listed shouldn't contain biosolids. This is another reason why it's important for you to understand what's in the amendments you're putting on your garden.

SALTS

Just as too much salt is bad for your health, it'll do your soil and plants no favors. Soluble salts can be found in high concentrations in some fertilizer products, composts, and manure. Salts can also make their way into your garden via runoff from de-icing materials and along coastlines. Salts in the soil are harmful because they can draw water away from plants' roots. Plants will take in some of the salts, and an excess of salt can cause the leaves to burn.

Salt concentration is another reason to use care to avoid over-fertilizing plants and over-amending soils. Regular rainfall can wash out some of the salinity. In fact, if a nutrient test shows your compost is high in salts, you can let it sit out for a few months to let the rain reduce the salt content.

COMPOST

Despite all of the commercial fertilizer and soil amendment products available to gardeners, it's hard to beat simple, decayed organic matter, also known as compost. Ingredients you can add to your home compost bin include kitchen scraps (except meat, dairy, and oils), shredded cardboard, brown paper bags, napkins, garden and yard cuttings, fallen leaves, and coffee grounds and teabags.

Compost, whether homemade or purchased from a reputable source, improves soil structure, adds nutrients to the soil, feeds the soil life, and boosts water-holding and water-filtration capacity. This is why compost is sometimes referred to as *black gold*.

In addition to the benefits compost offers to soil, consider the other ecological benefits:

- Food waste makes up nearly one-quarter of municipal solid waste, taking up vast landfill space itself. Organic waste, including food, yard waste, paper, wood, some fabrics, and more, makes up three-quarters of all waste.
- As organic materials decompose in an anaerobic environment, such as a landfill, they release carbon dioxide and harmful methane, both greenhouse gases.
- With a moisture content of 73 percent, food in landfills can pollute groundwater as its moisture leaches out of the landfill, carrying chemicals and heavy metals with it.

Compost can be homemade or purchased in bulk or bags.

There are several different types of compost that you can make at home and find for purchase.

Basic Mixed Compost

There are two ways to create basic mixed compost. One is hot composting, which involves just the right ratio of carbon-rich materials to nitrogen-rich materials, water, and oxygen. The other is cold composting, which involves putting the organic materials into a pile and walking away. Hot compost requires more attention and yields quicker results. In actuality, all but the most serious composters probably use a hybrid of these two methods. (See Efficient Ecological Composting on page 191 for more about making compost.)

For anyone making compost at home, there's a dizzying selection of compost bins available for purchase and at least as many DIY plans for building one. Select the bin type that makes the most sense for your situation. Gardeners with limited space might opt for a barrel tumbler, while those with the yard space to spare could go for a three-bin compost system.

If you're new to composting, you can probably take a workshop with your Cooperative Extension, local garden club, or municipality to boost your confidence in the process. If composting at home is not an option, you may find a composting service in your area, which would pick up your organic materials year-round and drop off finished compost a couple times a year.

Vermicomposting uses worms to break food scraps down into a rich soil amendment.

Vermicompost

Worms, specifically red wiggler worms, do much of the work in vermicomposting. (In Latin, *vermis* means "worm.") A vermicompost system starts with a dark, moist habitat for the worms, which could be a commercial setup or a DIY bin with small holes drilled in it for aeration. The worms need moist bedding, such as dried leaves or shredded newspaper or cardboard, and then you add food scraps and let the worms do their decomposition work.

From a vermicompost bin, you gain two useful substances:

- Worm castings are the nutrient-rich waste the worms leave behind. Rich in microorganisms and nutrients, worm castings are a great soil amendment.
- Worm leachate is the liquid that drains from the vermicompost system. Dilute this 10:1 with water, and use it to give your garden soil a boost.

Anyone can vermicompost in systems large and small. This is a good option for gardeners with limited outdoor space, as you can keep a vermicompost bin under your kitchen sink, in your garage, or on the back porch. These can also fly under the homeowner association's radar.

Bokashi Compost

For the organic materials that cannot be composted in basic and vermicompost systems, bokashi composting picks up the slack. Bokashi is a fermentation process that takes place in a sealed, anaerobic compost bin or bucket. Starting with a bran inoculated with specific microorganisms, you can add dairy, meat, oils, and animal wastes to a bokashi system for decomposition.

Similar to vermicompost, you're left with two nutrient-rich results:

- Bokashi compost is not-quite-finished compost. After breaking down in the bokashi bin for about two weeks, the resulting decomposed materials are acidic. You can bury this in soil in preparation for a future garden or add it to a basic compost bin to continue decomposing.
- Bokashi leachate, the liquid that drains from the system, can be diluted 1:100 with water for garden plants.

Buying Compost

The larger your garden, the more compost you'll use, and it's reasonable to think you may not be able to make all of the compost that you need. Similar to sourcing manure, you'll want to be sure your compost is free of herbicides. It's worth asking about the process used to make the compost, too, as you want pathogens and weed seeds to have been deactivated at high temperatures.

Efficient Ecological Composting

On the surface, composting is simply throwing organic matter in a pile and letting it break down; however, there is science behind it that makes the process most efficient and rewarding. Mostly, your job as a compost tender is to support the microbes that are doing the work of breaking down the organic matter.

A multi-bin composting system allows you to add new yard waste to one bin while the other bins are already in a state of decomposition.

Here are some tips:

- **Balance your carbon and nitrogen.** Organic materials are biologically complex, but their primary components are carbon and nitrogen. In "compost speak," brown materials have a lot of carbon, and green materials have a lot of nitrogen. Browns include dead leaves, cardboard, and wood chips. Greens are fresh vegetables, grass clippings, and coffee grounds. The ideal compost environment is 30 parts brown to 1 part green.

  Naturally, all organic materials have some carbon and some nitrogen, so it's impossible to give an exact recipe for perfectly balanced compost inputs. You can look up an analysis of the carbon and nitrogen content of each of your compost inputs to calculate the exact carbon-nitrogen ratio, or you can let your compost pile tell you what it needs. When your pile is sopping wet and smelly, add more carbon-heavy materials; when it's dry and seemingly lifeless, add nitrogen-rich materials.

- **Keep the pile aerated.** Some microbes thrive in an environment without oxygen (anerobic conditions), while others need oxygen to live, eat, and reproduce (aerobic conditions). Aerobic conditions make for a healthy compost pile. When a pile turns anaerobic, you'll notice bad odors and slimy, rotting organic materials rather than odorless compost with a reasonably moist consistency.

  How your pile becomes aerated depends on the compost system you're using. In a static bin, you can turn the compost with a pitchfork or compost turner. In a barrel-tumbler system, you simply turn the crank. More elaborate compost bins may include aeration pipes, though it's not likely you'd want to install one of those in a backyard setup.

- **Keep the pile moist.** Plunge your hand into the center of the compost pile and pull out a handful. It should feel like a damp sponge, not wet enough to wring out. Compost microbes do best at 40 to 60 percent moisture. In wet climates, a cover of some sort may be necessary so you can better control the amount of rain infiltrating the pile. In other places, you might have to add water during dry periods. Keep in mind, too, that food scraps are mostly water and will provide some moisture for the pile.

- **Support biodiversity.** Compost is another place where diversity is important. A compost made from various sources of organic material will provide a well-rounded soil amendment. Consider this in your own backyard compost bin and as you purchase compost from outside sources. At home, add kitchen scraps, yard waste, and whatever other organic materials you can source. Leave out the oils, meat, and dairy, as they'll attract rodents, cause an odor, and throw off the microbial activity.

Manure for Vegetable Gardens

A sign reading "Free horse manure" is a common sight in rural places. (Sometimes chicken replaces horse.) Farmers and backyard-animal keepers have a big chore in managing manure, and home gardeners can take advantage of this mutually beneficial situation.

While manure isn't the best choice for use on native plants due to its nutrient content, in vegetable gardens you can apply manure, fresh or composted, to enliven microbial activity in the soil, boost soil organic matter, and add nutrients. Manure from ruminant animals (such as cows and sheep), horses, and poultry is generally considered safe for garden use. That from pets and pigs is not safe, as it may contain parasites, in addition to harmful pathogens.

In addition to local manure sources, bagged manure and commercially composted manure are widely available. Using local manure is an ecological choice for a vegetable garden soil amendment or fertilizer. You're helping to recycle nutrients that might otherwise cause harmful runoff into waterways while creating a more closed-loop garden system by using nearby nutrient sources.

Be sure manures are fully—and safely—composted prior to use in vegetable gardens.

One important question to ask in sourcing local manures is related to herbicide use. If the animals grazed in pastures or on hay that was treated with herbicides, the herbicide can still be present in the manure (even after composting). These chemicals can interfere with both plant growth and soil life.

COMPOSTED VS. FRESH MANURE

You could think of manure as already-composted vegetative material with additional enzymes mixed in. Alongside those enzymes are pathogens that could potentially be harmful to people and waterways. There are nuances to using composted and fresh manure.

While there are varying degrees of "fresh" manure, here the term refers to any manure that hasn't undergone proper composting. Fresh manure is high in nitrogen, with the highest concentration in the freshest manure. While this is a benefit to depleted soils, this concentration can "burn" plants. Composting mellows out the nitrogen content. Both fresh and composted manure also contain phosphorus, potassium, and the lesser nutrients, so it's important to apply these only when your soil needs an overall recharge.

Fresh manure may also be high in weed seed content and in pathogens. Both are a concern to every gardener, and both can be neutralized by composting. The weed seeds in fresh manure can create a weedy mess as they germinate. Horse manure is notorious for carrying seeds. Weed seeds can be killed when the compost pile reaches at least 140°F (60°C).

Pathogens are a next-level concern. No one wants to work in the garden alongside a concentration of *Escherichia coli* or *Salmonella* bacteria, nor do you want to harvest contaminated produce and herbs. Composting manure at 131°F to 140°F (55°C to 60°C) for several weeks and regularly turning the pile to ensure all of the manure has been exposed to those high temperatures destroys the pathogens for safer use in an edible garden.

BAGGED, COMPOSTED MANURE

A convenient but less ecologically minded alternative to using local manure on your vegetable garden is bagged, composted manure, which comes packed in plastic. It's easier to manage than a pickup truck–load of fresh manure, you can be sure it doesn't contain residual herbicides, and the nutrient content is conveniently printed on the label. You can also add bagged, composted manure to your regular compost to augment what you've been able to make yourself.

APPLYING COMPOSTED MANURE

Composted manure can be used as you would regular compost. Spread a few inches over the soil in your vegetable garden and incorporate it into the top layer or use it as a top dressing. There's no concern about having to wait to harvest your produce.

Plant debris left to decompose in place eventually becomes a terrific soil improver. Here, you can see the old plant stems have been chopped into smaller pieces and left on the bed as mulch with fall leaves.

PLANT DEBRIS

Several times already you've read about leaving plants in place to decompose, cover the ground, and provide insect and wildlife habitat. Below ground, the roots left behind after plants die are also beneficial to the soil. As they break down, the roots' nutrients are deposited directly in the root zone for other plants and soil life. The space that the roots occupied become channels for air and water movement.

BIOCHAR

Used in Indigenous and traditional agriculture for thousands of years, this porous, carbon-rich substance is charcoal made from wood and agricultural byproducts. Biochar improves water-holding capacity, and while biochar itself isn't high in nutrients, it has nutrient-holding capacity, keeping more nutrients in the soil for a longer period.

You can make biochar in a homemade system, or you can purchase the finished product. Biochar made from different source materials has different beneficial properties, and not all biochar is suitable for all soils. Biochar can increase soil pH (make the soil more alkaline), so if you already have basic soil, this isn't the amendment for you.

MULCH *in the* ECOLOGICAL GARDEN

MULCH COMES IN MANY FORMS, both organic and synthetic. The characteristic that unites them all is that they cover the ground to mimic the ground cover that we see in nature: the stones, fallen leaves, twigs, and plant cover of the forests, fields, and deserts.

Mulch is a boon in any garden. Among the many reasons to cover the ground with mulch are:

- Preventing weed growth
- Cooling the soil during hot weather, which also slows evaporation
- Warming the soil in cool weather, which insulates plant roots so they withstand winters
- Holding moisture in the soil
- Holding the soil in place, preventing erosion in windy areas

Arborist chips make a useful mulch for perennial and shrub beds.

- Improving the soil as it decomposes (in the case of organic mulches)
- Reducing fungal and bacterial transfer from the soil surface to plant leaves by preventing raindrops from hitting the ground and splashing back onto the plant
- Providing habitat to beneficial insects
- Providing a consistent, attractive backdrop for your garden

TYPES OF MULCH

While you can find plenty of examples of synthetic mulch, including rubber chips and plastic sheeting, these are not ecologically minded choices. The mulch options outlined here are natural materials.

Whatever type of mulch you choose for your garden, know its source. Mulches that come from plants treated with herbicides or pesticides can hamper or altogether halt your garden's progress.

Straw. Used more often in annual vegetable gardens than in ornamental and perennial gardens, straw is the stalk remaining after harvesting a grain crop, such as wheat or oats. While it's inexpensive and widely available from farmers and at garden-supply stores, straw is messy to transport and decomposes (or blows away) faster than other mulches.

Wood chips. Wood chips are a common mulch for gardens, and especially for perennial gardens and shrub areas. In all areas with tree cover, wood chips can be found in bulk at garden-supply stores and sometimes for free from municipalities and arborists who are trying to dispose of this tree-trimming byproduct. Keeping this resource within the community is a good example of a closed-loop garden economy. Wood chips decompose slowly, are not likely to blow away, and hold onto moisture close to the soil surface. Woodland gardens are the best candidates for using this mulch; partially decomposed mulch is best, if possible.

Making Leaf Mold

Leaf mold is a simple ecological gardening task, and it takes care of several challenges you may face:

- It's a use for the fallen leaves that your neighbors or homeowners association may be pressuring you to rake up.
- It's a use for your neighbors' fallen leaves, so they don't end up in the landfill.
- It's a valuable mulch.
- It creates habitat for invertebrates.
- It's free.

Leaf mold is also easy to make (much more so than high-quality compost). Leaf mold essentially is compost, as fungi work to break down the dead leaves.

You can make leaf mold in place, right in your garden. In this way, you're providing great insulation over the winter for the plants and roots underneath. As the weather warms in the spring, you may need to clear away some of the leaves that haven't broken down yet so dormant plants can reinvigorate and those waiting just beneath the soil can emerge. If you live in a windy area, you might need to corral the leaves in place with a short fence so they don't blow away before they have a chance to do their job. You can remove the fence in the springtime, once the leaves have mostly decomposed.

Some gardeners will suggest shredding the leaves with a lawn mower to speed the decomposition process. Shredding them also risks shredding the insects that are taking up residence among the leaves and the eggs they deposited there. Check this practice against your ecological gardening principles to determine whether it's right for you.

Shredded bark. From pine, cypress, or hardwood trees, shredded bark is similar to wood chips in its sourcing and benefits. Avoid dyed bark products.

Leaf mold. Leaf mold is partially composted leaves, which are a big benefit to soil life. Trees take in nutrients and minerals from wherever their deep roots travel in the earth, and their leaves contain traces of these. As the leaves fall and decompose, they leave behind these valuable deposits.

Leaf mold decomposes quickly, and if it's dry, the leaves can blow away. But you can make your own at home with fallen leaves from your and your neighbors' trees. (See Making Leaf Mold above.)

Rocks. Rocks, stones, or pebbles are an ideal mulch for arid and fire-prone areas. This is a common mulch in xeriscape dryland gardens. Rocks, stones, and pebbles make a great mulch over sandy soil not prone to weed growth. On rich, loamy ground, keeping weeds from growing up among the stones is more labor intensive.

This natural mulch material works well for plants native to arid climates, as it doesn't hold moisture against the plant the way a wood mulch does. These plants aren't accustomed to moist conditions, which could cause fungal and rotting issues.

Living mulch. Also called green manure, living mulch is any low-growing ground cover that fills in spaces between plants. On the plus side, roots of the living mulch keep the soil microbiome active and hold soil in place, and the presence of living mulch deters many weeds.

WHEN TO MULCH (AND WHEN NOT TO)

Mulch is almost always a good choice, but there are guidelines for when and how to mulch for the most gain:

- When planting a tree, mulch over the root mound helps hold moisture until the roots are able to branch out and find their own water sources. Keep mulch 2 inches (5 cm) away from the bark of the tree to prevent damage to the trunk from moisture and rodents.
- Mulch immediately after weeding to cover the soil and prevent new weeds from germinating.
- Apply mulch before the ground freezes in the fall or after it thaws in the spring. Mulching while the ground is frozen traps the cold in the earth rather than warming the planting beds, and this may delay emergence of early spring plants.
- Mulch bare ground when the soil has the right amount of moisture. Waterlogged soil will remain too wet with mulch, and parched soil won't soak up the water it needs when topped with a dry mulch layer.

There are also a few instances when it's better to leave the ground uncovered. These are important to recognize in an ecological garden.

- Do not apply mulch if the mulch smells off. Wood chips should smell like fresh-cut wood or have no odor. Aerate mulches giving off sour odors before you use them.
- Be aware of your garden's insect population. If you have unwanted visitors, such as snails and slugs, they may be living in your mulch. This is especially a problem in wet weather. You might have to pull mulch away from your plants until your beneficial insects (and birds) move in to help control the pest population.
- Give space to our ground-dwelling pollinators. About 70 percent of native bee species nest in the ground. Give them access to this space by leaving some of your garden unmulched. Different bees like different types of ground (sloped and flat, sunny and shady), so offer up a few patches of bare earth for their habitats.

Remember to leave some areas of the garden unmulched to provide nesting sites for native bees such as this mining bee.

PLANT PROPAGATION

LEARNING TO PROPAGATE your own plants by dividing them or starting seeds can be a great way to make your gardening and landscaping more sustainable, cost-effective, and well-suited to your local conditions. It isn't necessary, of course; great nurseries are a wonderful resource and not everyone has the time to devote to propagating plants, but it can be a really useful skill to learn if it fits into your life.

COST

The first reason most people start growing their own plants is to save money. Seeds are incredibly inexpensive compared to full-grown plants, and if you already have plants, saving seeds from them or dividing them can be virtually free. If you have a big area to fill with plants, a limited budget, and some free time in your schedule, propagating your own can be a magical solution.

CARBON FOOTPRINT

Many plants, especially those produced for big-box store retailers, get shipped long distances. Heavy pots full of wet soil are expensive to ship, and trucks burn a lot of fossil fuels to get from wholesale growers to your local big-box store. Just as with food, buying local is a great way to reduce the carbon foot print of your shopping, and starting plants at home is about as local as it is possible to be.

PESTICIDE USE

Most commercial growers rely on pesticides to produce perfect-looking plants that will sell well. While you may understand that a few holes in the leaves of plants are not a bad thing and a sign that they are feeding the local ecosystem, most shoppers will refuse to buy anything that doesn't look absolutely perfect. This makes it difficult to run a nursery without using at least a small amount of pesticides. If you don't have a good local nursery that you trust to produce plants without using pesticides that will harm your local ecosystem when you plant them in your garden, growing at home can be the simplest way to ensure everything you plant is safe for everything that lives in your landscape.

Propagating your own plants from seeds or cuttings is not only fun, it's also an ecologically sound practice.

SHARING

One of the great things about propagating your own plants is that growing two dozen plants usually isn't much more work than growing just three or four, so it is easy to make extras that you can then share with friends and neighbors. Sharing plants is one of the best ways to help make a vital ecosystem not just in your own yard but in the landscapes of others in your community. Telling people to cut down an invasive species or judging them for their chemical-soaked lawn will usually just make enemies. Gifting them a beautiful native plant and telling them which butterflies it supports, on the other hand, will make new friends and habitat at the same time.

INDEPENDENCE

There is nothing worse than sitting down in the spring to order your favorite variety of tomato, the one you've grown for years and always does great for you, and discovering that it is no longer for sale. Or getting the email from your favorite local nursery telling you that the owners have finally decided to take a much-needed rest or retire, and are shutting the business down. The only sure way to know

A grow light system allows you to start your own vegetables, annuals, and herbs from seed. It's also useful for propagating certain perennials from seed.

that you'll have the plants you want each year is to grow your own, setting you free from the vagaries of business decisions that may discontinue the best tomato you've ever grown because not enough other people are also buying it.

ADAPTATION

One of the best reasons to propagate your own plants is the ability to get plants that are the best adapted to your local climate and conditions. We've touched on this subject before when discussing collecting local seeds and selecting for the traits you find most desirable.

GROWING FROM SEED

There are many different approaches to growing plants from seed, but there are two simple techniques that you can use to grow a range of plants for your garden and landscape.

Starting Seeds Outdoors

The simplest method to start seeds is to let nature provide most of what they need by sowing seeds in pots and setting them outside to germinate. Starting in pots outdoors is particularly effective for hardy perennials, trees, and shrubs. For some native perennials, starting from seed is essential because grown plants are challenging to source at nurseries, especially if you are growing locally collected seeds.

By planting seeds in pots, it makes it easy to avoid competition with weeds during the germination process. It also makes it easy to move the plants indoors if there is extreme weather, and generally gives you more control over the process than if you sowed directly in the ground. And keeping the pots outside means you don't have to worry about providing artificial light or hardening the plants off to conditions outside before planting them in the garden.

Many hardy perennials are easy to start from seed sown outdoors, providing the seeds with the perfect growing conditions and ideal timing. 

TIMING

The best time to sow outside is usually fall or early winter. Many hardy perennials, trees, and shrubs have seeds that will not germinate until they have been exposed to a cold period. Planting outside allows them to naturally germinate at just the right time in the spring. By planting outside in the fall or early winter, you can let nature give them the cold treatment they need, and then as temperatures warm in the spring, they'll germinate when conditions are right. This method is sometimes called *winter sowing*, and it can take a lot of the guesswork out of seed starting.

LOCATION

The best place to start seeds outside is a bright but sheltered spot. Against the east wall of a house or shed is a great spot as the plants will get morning sun but be protected from the hot afternoon sun, and the wall can help provide a little insulation against extreme temperatures in the winter. Ideally, position the plants where they are exposed to the rain, but if the best spot is under an eave, that works too; you'll just have to water them more frequently.

Winter sowing in plastic jugs is a handy way to start seeds of many hardy perennials, including these rudbeckias.

CONTAINERS

A variety of containers can be used for starting seeds outside, but the best options are pots that are at least 1 quart (1 L) or bigger. Smaller containers tend to dry out too quickly. In addition to traditional nursery pots, you can reuse food containers, plastic jugs, or pretty much anything else that will hold soil and can have drainage holes punched or drilled in. It is best not to cover the containers with any sort of clear plastic lid as these can quickly overheat on sunny days and harm the seedlings. It is, however, a great idea to get a sheet of hardware cloth or another fine wire screen or mesh and lay it over the tops of the containers to keep squirrels and other rodents from getting into the pots and eating your seeds.

SOIL

Fill your containers with any good potting soil or seed-starting media. A range of options will work, and peat-free is a great choice (see page 21) if you can find it. This is, however, not a good time to use garden soil or homemade compost, as they are likely to include weed seeds. Starting with new, sterile, weed-free potting mix will make the process a lot easier as you won't have to guess if any seedlings you see are the desired species or weeds that slipped in by mistake.

SOWING THE SEEDS

Some seeds need to be sowed with a thick cover of soil, as they will not germinate unless they are in total darkness. Other seeds are the opposite—they need light to germinate, and so won't sprout unless they are sowed right on the soil surface. You can look up the germination requirements of specific seeds, but often just looking at the seeds will tell you what they need. Very tiny seeds—like poppy seeds or smaller—often need light to germinate, and so should be sowed right on the soil surface and then just covered with the lightest dusting of soil to help them stay moist. For larger seeds, the rule of thumb is to cover them with a layer of soil equal to their width—so the bigger the seed, the deeper you plant them.

You can sow many seeds in the same pot. This is sometimes called the *community pot method.*

Sowing them together makes it easier to care for them when they are small, and then you can separate them out into individual containers as they get larger.

CARE

Once sowed, just check on your pots, watering as needed. Once you see seedlings germinating in the spring, keep them watered. Once they have produced one or two true leaves, it is time to separate them out into their own pots. To do this, gently slide all the soil out of the container and tease the individual seedlings apart. This may seem radical, and like you'll damage them, but plants are tough, and they'll handle it just fine. Plant each seedling in its own pot and place the pots in a shaded spot for a few days to recover, then put them in a sunny spot to grow. Once they've reached a good size, you can plant them out in your garden or share them with a friend.

Starting Seeds Indoors

The primary reason to start seeds indoors is to lengthen the growing season. If you live in an area with short summers, food crops like tomatoes and peppers may not have enough time to produce, so starting indoors allows you to get a good harvest before frost arrives. It can also be useful for annual plants to enjoy a longer period of bloom during the warm part of the year.

While for perennials it is not necessary to start seeds indoors, you may want to just to jump-start their growth and get them up to a big, beautiful size faster, rather than waiting for them to bulk up over multiple growing seasons if sown outdoors. For some perennials, starting from seed indoors is essential because their seeds may require a particular treatment or the seedlings may require extra-sensitive care that's best delivered in a controlled indoor environment.

While lengthening the warm season by starting seeds indoors in the late winter or early spring is most common, you can also use indoor seed starting to extend the cool season. If you have a basement or other room that stays cool in the summer, you can use it to start fall plantings of cool-season crops like lettuce, which hate hot summer temperatures.

Pretreatment

If starting perennial seeds indoors, remember that they may need a cold treatment before they will germinate. A quick search online can help you find the specific treatment a particular seed needs. If it does need a cold treatment, you can provide it artificially by placing the seeds in a baggie with moist sphagnum moss or sand and putting the bag in the refrigerator. Another easy way is to sow them outside in the fall to get their cold treatment, but then bring them indoors in late winter to encourage them to sprout early for a jump-start on the growing season.

For seed starting indoors, your lighting setup can be a simple tabletop grow light or a multitiered shelving system like this one.

Lights

The most important part of starting seeds indoors is giving them enough light. Windowsills, even the sunniest of them, simply do not provide enough light for healthy seedlings, so you will need to use artificial lights unless you are lucky enough to have a greenhouse. Luckily, LED lights now are extremely energy efficient, affordable, and produce great-quality light for photosynthesis. You can purchase special LED grow lights, specifically designed to produce the wavelengths that plants use the most for photosynthesis, but that is not at all needed for seed starting. Any LED bulb marked as "cool white" will produce a great quality of light for seedlings. A good, simple solution is to buy two LED shop lights available at any hardware store and hang them next to each other.

Soil blocking is a seed-starting method that requires no containers. Instead, seed-starting mix is compressed into "blocks," which are then planted with seeds. Soil-blocking devices come in various sizes and are easy to use. The blocks are watered from the bottom to avoid disintegration.

You don't have to be fancy when starting seeds. However, do pay attention to the germination requirements for each specific seed and use sterile potting mix formulated specifically for seed starting.

Containers

When starting indoors, you can either sow one or two seeds in each individual pot or use the community pot method as described earlier for starting seeds outdoors. Sowing in individual pots makes less transplanting work, but it is a little less efficient when you have a small number of seeds. Because not every seed will germinate, you may want to sow two seeds per pot to ensure no pots are empty. Any extra seedlings will need to be pulled out once they sprout.

In either case, you can use standard nursery pots or repurpose other containers. If you are starting a lot of seeds, using standard nursery pots is a lot easier because their standard size means they will dry out at the same time, making watering a more regular, predictable chore.

You can sow individual seeds directly into large pots, but it is usually better to start in small individual containers and then pot the seedlings up into bigger ones as they grow, simply because it takes up less space under your lights. Small plants in small pots can be clustered together under just a few lights, using less electricity, and then you can turn on more lights as needed as you pot up into bigger containers.

Timing

Knowing when to start sowing seeds indoors is critical. For vegetables, if you start too late, they won't be big enough when it is time to plant them out in the garden. Sow too early, and they'll get too big to fit under your lights. For perennials, it's more about ensuring the seedlings have enough time to develop a good root system before the plants are transplanted outdoors. Most perennials require a longer germination and growth period than annuals and vegetables. Consulting a resource specific to growing perennials from seed is helpful in getting the timing right.

For vegetables, the exact timing will differ for each plant, but it is usually calculated in terms of weeks before your last spring frost. In early winter, gather all the seeds you want to start, look up their timings and your last frost date, and mark on the calendar the day they need to be planted to get it just right. It is also useful to keep records on how they performed. Plants will grow at different rates depending on temperature, so if your growing space runs cool, they may need to be started a week or two earlier, and if it runs hot, they might be better off started a little later. Good recordkeeping each year will help you to perfect your sowing schedule to get it right on the mark in the future.

HARDENING OFF

Indoor growing conditions are much different than outdoors, where there is blustering wind and, most critically, damaging ultraviolet light in the sun's rays. If you take seedlings that have been growing indoors and put them straight out into full sun, they are likely to burn and sun scald, showing patches of brown or bleached-out sections on the leaves. Hardening off is the process of slowly adapting plants to outdoor conditions. The usual recommendation is to start the process a few weeks before you want to plant the seedlings out, starting by just putting them outside for an hour, then the next day two hours, and so on until they can be outdoors all day.

Hardening off can also be done less formally by taking advantage of any warm days you get and putting seedlings outdoors as much as you can. Through much of indoor seed-starting season, there will be the occasional mild day when you can move your seedlings outside and let them adapt to outdoor conditions. If you do this regularly from the time they are small, they will mature fully adapted to bright sun and wind, and will transition to the outdoor world more smoothly when it comes time to plant.

DIVIDING

Dividing perennials is one of the most straightforward ways to multiply your plants. In the simplest terms, all it means is digging a plant up and cutting or slicing it apart into multiple pieces.

The key point to understanding perennial division is to focus on the crown of the plant—that's where the roots and the shoots meet, usually just at or just below the soil surface. At the crown of a plant, you'll have many roots emerging, and usually several different shoots coming up. When dividing a plant, as long as each piece you create has some roots and shoots attached, it can grow into its own new plant.

One sign that it's time to divide a perennial is when it develops the shape of a donut with an empty or sparse center.

Timing

Usually the best time to divide a perennial is in the spring right before it pushes out new growth. At that stage, you should be able to see large buds beginning to grow that will expand into new shoots for the year. Make sure there is at least one bud on each division. Dividing in the spring is usually the best for a few reasons. First, as the plant goes into active growth in the new season, it will be able to recover from the division and grow new roots to replace those lost while digging it up. It is also usually easier to handle a plant before it has grown a lot of leaves in the spring.

However, there are some plants that are best divided at other times of the year. Peonies, for example, only grow new roots in the fall, so late summer division is best. When in doubt, checking references about your specific plant will give you guidance on the best time to divide, but most perennials can be divided in spring or fall.

A sharp knife or shovel is useful for dividing perennial clumps.

Digging up

When digging up a plant to divide it, your focus should be on keeping as many roots as possible, which means lifting quite a large rootball. You'll never be able to get every root, but the more you are able to save, the better. One strategy to preserve a lot of roots is to focus on lifting and prying the plant out of the soil as much as possible, rather than slicing through the rootball with a shovel. In light, sandy soils, you can use a garden fork inserted into the ground around the crown and lifted up to pull the root mass from the soil. In heavier clay soils, you may need to use a shovel, but again, focus on prying the plant up carefully with the shovel rather than slicing around the whole plant.

Separating

Once your perennial is out of the ground, try to knock as much soil off of the roots as you can to make it easier to handle, lay it on its side in a cool, shaded spot, and then start the process of dividing it. There are a few techniques you can use. The first is to go in with your hands and try to pull the clump in half, but usually you'll need to work a trowel, garden fork, or shovel through the crown to get it to pull apart. Different plants will respond differently. Some, like bee balm and coneflowers, generally pull apart very easily. Others, like many native grasses, have dense, woody crowns that may need to be cut apart with a pruning saw. Pulling apart is always preferable to cutting, if possible, as it will save more of the roots and do less damage to the crown. If you can't pull it apart, the next option is to slice a few inches into the crown, and then try pulling again. Usually just cutting halfway through will be enough to get things started.

How many sections you divide the mother plant into is up to you. In theory, every piece that has some roots and at least one shoot or bud will grow. If you just cut a plant in half, you'll get just two new plants, but they'll be big and mature looking right away. If you carefully cut apart every single piece to just one single shoot or bud with some roots, you'll get many more plants, but each one will be very small and may take a year or two to bulk up and look mature again. It all depends on your goals and how many new plants you want or need.

Replanting

After all the sections have been cut apart, it is time to replant. They can go straight into the ground in your garden or you can pot them up into containers if you want to give them to friends or don't have a spot for them right away. After replanting them, be sure to water deeply, and keep them well watered for the next couple of months while they recover from the stress of division and regrow their root system.

THE NATURE *of* WORK

THERE IS MUCH WORK to be done in any garden. Keeping the focus on the ecological nature of this work adds a nuance to gardening that conventional gardening, focused on the human element, may overlook. Preparing your space, maintaining the garden with insects and wildlife in mind, doing your part to create and maintain healthy soil, and propagating plants for yourself and others are all important ecological gardening techniques to practice.

In the next chapter, you'll take the principles of ecological gardening you've learned throughout this book and apply them to other areas of your life. It's here that the holistic picture of how these pieces connect becomes obvious.

Propagating your own plants means you'll save money, control the inputs used to grow your plants, and generate a thriving ecologically sound garden full of native plants you've grown yourself.

CHAPTER 7

LIVING LIGHTLY

WHILE YOU'VE DELVED into the larger impact gardening has on your well-being and that of your community and natural surroundings, you don't have to dig too deep to see how your gardening practices are also connected to other practices in your life. As you begin to recognize ways you can garden more in harmony with the world around you, you start to see these opportunities everywhere. And you may be surprised by how these opportunities are interconnected.

Everyone is on their own journey through the ecological landscape. Each person is navigating cultural beliefs and an economic framework of what it means to exist in the United States today. So much of operating as a *part* of natural systems, rather than as something separate from natural systems, requires a mindset shift. You will find your way, just as everyone around you will find theirs, and the journey won't look the same for everyone.

In this chapter, you'll learn about taking your ecological gardening approach another step further. We'll look at reducing plastic use, finding alternatives to gas-powered outdoor equipment, and working toward a closed-loop garden system.

A Circular ECONOMY

THE MODERN CAPITALIST society operates using a linear economy. We make money, we buy a thing, we use that thing, and we throw the thing away. Now we need more money to buy more things that will, in turn, become waste.

A circular economy bends that one-way track so that it feeds itself. It's a cycle of services and consumption that includes sharing, reusing, repurposing, and recycling as much as possible. In a truly closed-loop circular economy, there is no waste.

Nature is a closed-loop circular economy. If humans were to disappear from the planet, the trees would still grow, birds would still sing, and fish would still swim. While we are a part of nature, we don't control nature's processes: The cycle of life and death creates the conditions necessary for the natural world to function. (You read about the nutrient cycle in chapter 6.)

Mimicking nature's circular economy in a modern garden is aspirational yet possible. A closed-loop system isn't a solitary system. As nature is a whole community of organisms, so is your garden. You know how to harness the nutrient cycle. There are also other ways you can work in your own space and reach out to those around you to help bring around this circular economy.

SOIL BUILDING

After reading about soil building, composting, and the nutrient cycle, you may have your own ideas about how your soil-building effort can be part of a closed-loop system. Unless you have a lot of space and organic materials on hand to make your own compost, you'll probably have to bring in some soil fertility from outside your yard. Look for resources in your own community, such as a neighboring chicken keeper who can share the contents of their coop, or a microbrewery that would like to offload a batch of spent grain—both of which you can add to your compost at home. Your municipality might also offer free compost as part of their waste-management plan. The closer the source, the better.

By keeping resources, such as wood chips and grass clippings, on site and using them as mulch creates a closed-loop system where little goes to waste.

FOOD WASTE

For vegetable gardeners, food waste is vexing. After you've gone through the time and effort of growing this food, having to throw it away can be frustrating. There are a lot of ways you can turn food "waste" into an asset, closing this part of the loop. Here are a few examples:

- Salvage vegetables and fruits that are past their prime by way of sauces, purees, and jams. As long as they don't show signs of rotting or molding, a bowl of tomatoes too soft to slice for sandwiches can still make a sauce, for example.
- Freeze what you can't use fresh. Most vegetables can be blanched (quickly boiled and cooled) and frozen for future use, and fruits usually don't even require the blanching step.
- Keep a bag of vegetable peelings and scraps in a container in the freezer, and use these to make stock. Use this stock in recipes rather than purchasing ready-made stock in cans and plastic pouches. The tops and ends of carrots, garlic and onion skins, leftover bits of herbs, and squash peels are a few produce items that make great stock ingredients.
- Share your harvest. An abundantly producing garden can be a gift to your neighbors as well as your own family. Don't waste this opportunity to keep up your relationship-building efforts.
- Let food-waste reduction be part of your ecological garden design from the start. If you know your family won't eat kale, either don't plant the red kale or plan ahead to give away this kale to neighbors and friends.
- Return the rest to the compost bin. If the produce you grew can't feed you, let it decompose, and return it to the garden so it can feed the soil.

Composting keeps food waste out of landfills and generates a useful soil amendment on-site.

ON-SITE PROPAGATION

A closed-loop system isn't meant to stifle garden abundance or creativity; if anything, engaging with this principle opens you to the abundance that is already within your ecosystem. While trips to the nursery are outside of the loop, propagating plants and savings seeds are well within it. Propagate plants and save seeds from your own plants, and exchange these with others. You'll keep your garden inputs closer to home and build reciprocal relationships with other gardeners.

REUSE AND REPURPOSE

What if there were no such thing as a landfill, and you had to find a way to re-employ everything that came into your possession? You would become more selective about what you brought into your life and more resourceful about how to keep it in use. Practice that in your garden. This concept will come up repeatedly in this chapter.

REPAIR

As an item becomes worn out, the easiest thing to do is to get rid of it and get a new one. The most ecological thing to do is to repair it.

You've likely heard someone older than you remark, "They just don't make things like they used to." It's true that manufacturing processes have introduced more inexpensive, poorly made goods into the world, and these are quicker to wear out and harder to repair than their high-quality counterparts of the past. This fact is a push in the direction of investing in higher-quality items, even if that means you can afford fewer of them.

Repairing items was a skill our grandparents had. Mending clothing, reinforcing tools, and repairing gardening equipment are less common skills today; however, these skills are far from lost. There are many ways for you to brush up on your ability to make repairs:

- **Tap into your community.** Among all of the relationships you've built, who already has these skills? Look especially at intergenerational collaboration. Sharpening pruning tools or replacing a broken shovel handle are skills you may be able to learn from others.
- **Read a book.** While repairs for electronic and mechanical items have become more complicated over time, basic hand-tool repair, fencing reinforcement, raised-bed fixing, and similar work hasn't changed. New books are nice, and older volumes will still teach you what you need to know.
- **Don't forget the internet.** Tutorials and videos abound for how to repair everything from garden hoses and flowerpots to rubber gardening boots.

Keep your garden tools in shape by regularly cleaning and sharpening them. Doing so extends their life.

SHARING RESOURCES

The idea that we should share anything is rooted in who we are as humans. We evolved to cooperate in communities. Sharing resources is a hallmark of truly ecological gardening for several reasons: Sharing means you have to purchase fewer items, which reduces the ecological footprint of your garden work; by investing with others, your purchasing power can go farther, meaning you can source better-quality, longer-lasting equipment; and by keeping resources within your community, you're closing the economic loop.

Shareable garden-related resources include things like:

- Tools you don't use every day: soil blockers, lawn mowers, wheelbarrows, post-hole diggers
- Bulk purchases: compost, mulch
- Knowledge and labor: books, subscriptions, workshops, workdays
- Kitchen equipment: canning equipment, a food dehydrator

Practice sharing resources outside of the garden as well. In addition to going in on purchases with people in your ecological gardening community, you can find shared equipment through lending libraries. Your own city or county library might have a "library of things" in addition to their book collection, or your community could have a stand-alone nonprofit tool library that lends all manner of items for a small membership fee.

BUYING LOCAL

Throughout this book, several concepts in setting up an ecological garden have reflected the importance of buying local. The idea of a circular economy ties this all together in a bow. From seeds to tools to lumber for raised beds, the benefits of locally sourced, locally crafted, and locally sold goods make "Buy Local" more than just a slogan.

Reduce shipping. There are many reasons to slow the transport of goods. Let's start with how your item gets to your home. First, it's manufactured in a country that's most likely not the United States. It travels here from overseas in a bulk shipment. The shipment is divided up at a central warehouse, which covers a dozen or more acres, and sent in bulk shipments to warehouses around the country, which have also paved over acres of once-open space. You click "buy now," and your individual product is individually padded and packaged. It's then sent, probably by airplane and several trucks, to your door.

If you had purchased this item locally, you would save the individual shipping padding and packaging, the fuel involved in the transport to your door, and the potentially low-wage and unfavorable working conditions inherent in multiple parts of this chain.

A note about accessibility: It's important to acknowledge the good and bad in every situation. For people who are unable to or find it difficult to leave their house for shopping, shipping goods directly to their door has become a lifeline. Nothing in this book is meant to be a blanket statement against strides made to make basic goods and services available to all, just to ask you to think about how the choices that you're making align with your ecological thinking.

Buy your plants from local growers whenever possible. Ask questions about how the plants are grown, where the seed was collected, and whether the plants are local to your region before purchasing.

Buying Local Food

While using your purchasing power to support local businesses, look at your food-buying habits as well. Many of the vegetables, fruits, and herbs you're not growing yourself can come from your region in season. In addition to all of the reasons listed for buying local goods and services, local food has additional benefits:

Farmers keep open space open. You're aware of the ecosystem services that an ecological garden provides. Multiply this by a farm scale, and you understand the difference that even small farms can make. In a time when land is being developed at breakneck pace—California alone is losing 40,000 acres (162 square km) of agricultural land per year—the people working to keep farmland productive deserve support.

Local food is more nutritious. For the most part, nutrition declines from the time the plant is cut, so when you consume vegetables harvested yesterday from a nearby farm, you're getting more nutrition than from those that were harvested a week ago and shipped across the country.

Local food tastes better. Big agricultural producers who sell to grocery stores and wholesalers have to grow produce that is uniform in size, shape, and color and holds up to handling and shipping. That produce doesn't always have a lot of flavor. Smaller producers selling to local markets have the freedom to grow better-tasting varieties of fruits and vegetables that may not have the same shelf life and appearance.

You know your farmer. You're getting to know the farmer's practices at the same time that you're getting to know the farmer. When you're able to meet the person who grows your food, you can ask about whether they use chemicals, who provides labor on the farm, and more. You may even find someone to trade gardening tips with.

Shop your local farmers market to keep your dollars close to home and support local growers.

Farmers markets support communities. Farmers markets are not the only way to access locally grown food, but if this is where you choose to purchase your local food, know that they provide additional community services. Across the country, farmers markets fill a void in areas of food apartheid, sometimes called *food deserts*, where fresh foods aren't available. They often offer kids educational programming; incentives for people shopping with Supplemental Nutrition Assistance Program (SNAP) cards, formerly known as food stamps; a fun community gathering space; recipes and chef demonstrations for preparing healthy foods; and other activities.

Spend local to stay local. Locally owned businesses, including garden centers, plant nurseries, and farmers markets, keep more money in the community than chain stores: For every $100 you spend at an independent retailer, $52.90 is recirculated locally, versus $13.60 out of every $100 by chain retailers. Business profits are reinvested in your community—in the form of wages, by purchasing goods and services from other businesses, and as charitable support—rather than reinvested in some city far away. More business income means more tax revenue, which could mean better public infrastructure, including roads, schools, libraries, and recreation. This is called the *local multiplier effect.*

Support your community's efforts. Independent businesses make charitable contributions, but they also support employees in volunteer participation and help guide organizations as members of boards of directors and advisory committees. One study in Maine showed 91 percent of local business owners contribute to their community in some way.

Create jobs for your neighbors. Two out of every three new jobs are created by small, local businesses. These businesses provide more than 52 percent of jobs in the United States.

Make your community stand apart. Consider the "Keep Austin Weird" slogan, coined in 2000 to celebrate the uniqueness of this Texas city. If Austin had only the same chain restaurants and big-box stores as every other city with no independent businesses, they wouldn't have much to talk about. Interesting communities retain and attract residents and tourists. Strong commercial districts give communities gathering spaces. Keeping independent businesses alive keeps the spirit of a place alive, adds character, and gives residents something to be proud of.

Know your neighbors. Here it is again: the importance of relationships in an ecological garden (and lifestyle). In general, you'll receive better customer service and greater know-how from employees at local establishments, garden-related and otherwise.

Let's say your community no longer has a local nursery, but you do have a big-box hardware store—a reality for many today. Purchasing from this store directly is still a better option than shopping online for many of the reasons listed here.

THE ZERO WASTE MOVEMENT

The Zero Waste International Alliance defines zero waste as "the conservation of all resources by means of responsible production, consumption, reuse, and recovery of all products, packaging, and materials without burning them and with no discharges to land, water, or air that threaten the environment or human health."

This is a tall order, and it requires more than your individual action. The zero-waste movement asks for product and packaging redesign with nontoxic materials, innovations in reuse and recycling options, and incentives for companies and people to change their lifestyles and habits.

As a result of and alongside this movement, cities and states have adopted waste-reduction goals, enacted laws requiring food waste be composted and banning single-use plastics, and offering incentives for residents and businesses to change their practices. Corporations, too, are making strides toward zero waste. There are likely many examples of this work, even in your own community. Zero waste is a great goal to have in the garden too.

Recycle garden debris by composting, collect and use rainwater on-site, opt for a plastic-free landscape, and give unused gardening tools that may have otherwise found their way to a landfill to a local community garden or tool-lending library.

RESOURCE CONSUMPTION

Water and electricity are two resources whose use many of us can do a better job of regulating. Reducing use of these resources saves you money in addition to reducing environmental impacts.

Preserve water by carefully monitoring and targeting its use.

Do Away with Light Pollution in the Landscape

Since the Industrial Revolution, light pollution—man-made light shining into the sky at night—has increased each year.

Light pollution is a major disruptor of wildlife activity. It negatively impacts night-flying insects, including fireflies, moths, and many other nocturnal pollinators, whose attraction to artificial light causes disorientation and increased predation. Light pollution is also a detriment to birds, disrupting their migration patterns and increasing the chances of window strikes and vehicle collisions. And nocturnal animals such as bats and owls require darkness for hunting and navigation.

In addition, diurnal animals including toads, frogs, sea turtles, and salmon are impacted by light pollution. Whether it's disorientation or disruption of their natural activities, light pollution's impacts on wildlife are huge.

To limit light pollution, turn off outside lights completely or put them on timers or motion sensors. Replace white-light bulbs with yellow light. Add hoods or shields to your lights so they only shine down toward the ground, rather than up into the sky.

Light pollution affects many nocturnal pollinators and other insects. Turn off outdoor lights completely, put them on timers, or use motion sensors.

Water use. Sacred to Indigenous communities and essential to all life on Earth, water is an important enough resource that it warrants its own chapter in this book. (Refer to chapter 5 for all you need to know about water in an ecological garden.)

Spend time thinking about how you can reduce water use in and out of the garden, both directly (in your watering can) and indirectly (in the products you purchase, or don't purchase). Some ideas include:

- Install water-efficient irrigation and keep it maintained to avoid water waste. Refer to chapter 5 for more about irrigation.
- Fix leaks. This is an obvious way to reduce your water use. Across the United States annually, leaks waste as much as 900 billion gallons (3 trillion L) of water, which could provide for the water needs of nearly 11 million households.
- Don't waste food. Every plant and animal requires water to grow. Every bit of wasted food is also wasted water.
- Learn about how and where foods are grown. Food items have complicated water-use backgrounds that you would rarely consider while standing in the grocery store aisle. For example, California produces more than one-third of the vegetables and more than three-quarters of the fruits and nuts grown in the United States. Its year-round growing climate makes it ideal for agricultural production. Large swaths of the state often experience drought, which causes farmers to lean on irrigation. Because of reduced surface water availability, they've turned to pumping groundwater, which is damaging wells and aquifers.

Trace back your favorite foods to their source, and question their water use along that chain. For example, alfalfa tops the list of the thirstiest crops in California. While this isn't a plant that humans eat, it's grown for beef and dairy cattle feed (plus horse feed and human supplements). Beef cattle may drink up to 30 gallons (114 L) of water per day, and dairy cattle as much as 50 gallons (189 L)—or more, in extreme heat. You also need to factor in the water used to grow and process their other feedstuffs. Connecting the dots in how your food got to your fridge can have an impact on your water footprint.

- Reduce your overall consumption. Water is required in every manufacturing process. Reducing your overall consumption automatically reduces your water footprint.

Energy use. Whether coming from fossil fuel or renewable sources, energy production is an ecologically intense activity. Lessening your energy use in your gardening activities (and elsewhere) saves on utility bills while lightening your impact on your immediate community and the wider natural world.

A few of the following ideas will sound familiar, but here are some ways to reduce your energy consumption:

- Install energy-efficient lightbulbs, fixtures, and appliances, indoors and out. LED lightbulbs, programmable thermostats, programmable smart-home systems, and ENERGY STAR-rated appliances can significantly reduce your electrical usage.
- Turn off and unplug items not in use.
- Green up your laundry (and dishwasher). Wash your clothing in cold water. Let clothing and dishes air dry instead of running the dryer/heat-dry dishwasher setting.
- Harness heat gain. Uncover south-facing windows during the winter to let in as much sunlight and warmth as possible. In the summer, close those blinds and curtains to keep sunlight out.
- Seal your home. Drafty windows and doors and poorly insulated structures cause you to use more energy than necessary to maintain climate control.
- Use renewable energy. At home, reduce your dependence on fossil fuels by installing rooftop solar panels to harness solar energy or by replacing your heating system with a heat pump to take advantage of geothermal energy from the ground. Renewable-energy infrastructure may be eligible for tax rebates and incentives as well.

Aim outdoor lighting in a downward direction and use solar-powered lights for walkways and outdoor living spaces.

If installing your own renewable-energy infrastructure doesn't make sense, you might be able to draw your electricity from renewable sources through your electric-utility provider for a premium. Wind, solar, and hydroelectric are all renewable sources being harnessed at the industrial level.

The infrastructure and equipment involved with renewable energy carries with them their own ecological considerations. From the rivers dammed for hydroelectric to the wind turbines' potential impact on migratory birds and the rare minerals used in solar panels and batteries, account for the true cost of using renewable energy in establishing your ecological priorities.

GOING *Plastic* FREE

Reducing plastic waste in the garden is challenging, but not impossible. Start by opting not to purchase mulches and soils bagged in plastic.

Purchasing seedlings, rather than growing your own, often generates a lot of empty plastic pots. Many nurseries take back empty pots for reuse. Or, next season, consider growing your own plants from seed or cuttings in those used pots. Just be sure to wash them first.

TAKING A LOOK AT YOUR GARDEN supplies and equipment, you might be surprised by how much plastic is involved in modern gardening. We have plant pots, seed packets, weed barrier and mulch, soil and compost bags, wheelbarrows . . . the list goes on. The annual plastic use in North America is the highest in the world, representing 21 percent of plastic worldwide.

Annual global plastic production has doubled since the start of the century. Without plastic, modern life would be entirely different, which is to say that not all plastic use is bad; however, our overreliance on plastic and view of it as disposable is problematic.

There are several compelling reasons to reduce plastic reliance, in the garden and elsewhere.

- Plastic is a synthetic product derived from petrochemicals. The plastic manufacturing process gobbles up 4 percent of the world's fossil fuel use, and these processes are ecologically detrimental. While you might not see the direct effects of industry in your neighborhood, millions of people around the world suffer the health and environmental effects of related air and water pollution.

- Plastic creates a colossal waste problem. About one-quarter of plastic waste is "mismanaged," meaning it becomes litter. As much as 2 tons (1.8 metric tons) of plastic ends up in the ocean each year, and one estimate suggests that if current plastic use and disposal practices continue, the ocean could contain more plastic than fish, by weight, come 2050.

- Once plastic is here, it's here for a very long time. Some, but not all, plastic can be recycled. (See Not All Plastic Is Equal on page 218.) Disposable diapers are one example of a plastic product that cannot be recycled. If these end up in the ocean, it can take 450 years for them to degrade. As plastics break down—in the landfill, as litter on the side of the road, or in a body of water—tiny microplastic particles and even smaller nanoplastic particles can blow through the air, lodge in soils, and eventually break down into gases, including ozone-harming carbon dioxide and methane.

Despite these facts, you'll be hard-pressed to rid your garden lifestyle of *all* plastic. The good news is that opportunities abound for you to reduce plastic use and prevent unnecessary plastic from coming into your garden. A lot of these actions involve changing the way you think about consumption. Some of these actions are easier than others.

TAKE A PLASTIC INVENTORY

The more you're conscious of your plastic use, the easier it will be to do something about it. Start by writing a list of plastic in your garden collection. A few of these items were mentioned above. Don't forget to include things made from fiberglass, vinyl, PVC, and polypropylene. This exercise is not meant to frustrate you; rather, it's meant to create an outline for making plastic substitutions.

With your list of plastics in progress, note nonplastic alternatives for each. Here is a short list to get you started:

- **Seed-starting trays.** One alternative to plastic trays is to save your toilet paper rolls throughout the year and use them as small pots to start seeds. Another option is to do away with individual seed-starting containers and instead use soil blocks. Invest in metal soil blockers, which are handheld contraptions that press your soil mix into compact blocks. These tools will last the life of your garden.
- **Plant pots.** Use terra cotta, bamboo, newspaper pots, or cellulose fiber pots. If you're using the soil-block seed-starting method above, purchase a 4-inch (10 cm) soil blocker to make larger blocks to "up-pot" your seedlings.
- **Seed packets.** Source and trade seeds in paper packets. Paper is better for seed longevity as well.
- **Garden hose.** Rubber production has its own set of concerns, but it's a plastic alternative derived from a natural source.
- **Landscape fabric.** Use burlap, cardboard, or biodegradable-paper weed barrier.
- **Sheet mulch.** Use organic mulches. They may not have the same heat-trapping and weed-smothering capability, but they are an alternative to plastic.

Grow seedlings in soil blocks (see caption on page 202) to reduce plastic use when seed starting.

- **Soil and compost bags.** Purchase soil and compost in bulk. If you only need a small amount of soil or compost, bring your own burlap bag to the garden center and fill it from their bulk area. If you can use, store, or share a larger amount, have it delivered by the truckload.
- **Wheelbarrow.** At the end of its life, replace a plastic wheelbarrow with a metal one.
- **Floating row cover.** Row covers look like fabric, but it's typically woven polypropylene. Look for 100 percent cotton row covers.
- **Greenhouse plastic sheeting.** Invest in a glass greenhouse with a metal frame.
- **Watering cans.** Go back to using the old-fashioned metal cans.

Not All Plastic Is Equal

There are many different types of plastic. Some are recyclable and some are not. Often this distinction depends on the recycler used by your municipality (if your municipality even offers a recycling service). Some plastics are meant to last, and some are single use. Some "plastics" are not plastics at all.

When you must purchase plastic, reach for high-quality, durable goods, as well as those made from recycled content and those that can be recycled at the end of their lives.

DURABLE PLASTICS

If you must make a plastics purchase, make it a good one. Take seed trays as an example: You can find flimsy seed trays that will not make it through the season intact, and you can find durable, injection-molded seed trays that are meant to last.

COMPOSTABLE OR BIOPLASTICS

Over the past few decades, bioplastics have come on the scene. These are made from byproducts of corn or sugar-cane processing, forestry residues, and other organic materials that are mixed with chemicals to form a plastic structure.

Whether compostable plastics and bioplastics are actually better from an ecological standpoint depends on your perspective. If you're swapping out one single-use item for another, you're still working with materials that are not meant to last. Compostable plastics and bioplastics require water, chemicals, and energy in their manufacturing processes, and there are questions about whether these are any easier on the environment than those used in plastic processing.

Compostable plastics are meant to be composted in commercial facilities. If you were to put a compostable to-go coffee cup into your home compost bin, you'll find it's still largely intact after all the other organic materials have decomposed. Your home compost bin is not likely to reach the temperature or other conditions necessary to degrade the fibers of compostable plastics.

UNDERSTAND THE SYMBOL

Many plastic items bear the universal recycling symbol—the three arrows folded into a triangle, representing a Mobius loop. The number inside the symbol indicates the resin identification code, disclosing the type of plastic that item contains. There are seven main types of plastic, and each has a different recycling process. The decision behind which processes to use is usually a financial one. The recycler has to have a market for the end materials.

It used to be that this code was all you needed in order to know whether your recycling company accepts the plastic. It's now a less reliable indicator of recyclability, because plastic shapes and manufacturing techniques are so varied. Recyclers are now more likely to ask you to look at the code in addition to the shape of the item before you toss it into the bin.

All compostable plastics are bioplastics, but not all bioplastics are compostable. Some bioplastics are recyclable, but any that are marked as compostable cannot be recycled. You can identify a bioplastic by its recycling symbol: a 7 inside the triangle with PLA printed below.

Your best option for sorting through this mess is to contact your municipality or waste-handling company and get a list of which items go in which disposal bin.

If you must purchase plastic items for gardening, choose durable options that will last for years instead of flimsy alternatives.

- **Liquid fertilizer and amendment bottles.** You don't have a lot of alternatives here. It's possible but unlikely that your local garden-supply store would have these in bulk so you could fill your own container. An alternative is to purchase a large plastic container to share with other gardeners.

- **Drip irrigation tubes and fittings.** Here, too, plastic is the option if you're choosing to run irrigation to your plants. Consider how this practice fits into your definition of ecological gardening and whether it's something you want to continue.

- **Composite wood fencing.** Wood, metal, and concrete are all attractive alternatives.

- **Garden décor.** Opt for wood and metal alternatives, including handcrafted items purchased from local artisans.

INVEST IN A PLASTIC-LESS FUTURE

You now have an idea of how plastic fits into your garden life. The next step is to decide what to do about it.

The "reduce, reuse, recycle" motto came about in the 1970s. At the time, these ideas were novel. Today, the motto has little impact—we know the drill. But do we really know the drill? At this point, many people have fallen into complacency.

There are some macro-level actions that you can take to "reduce, reuse, recycle" into action in the garden:

Break up with cheap goods. One reason plastic is now ubiquitous in our lives is its low economic cost, both as a raw material for manufacturers and as finished goods for consumers. When you factor in the ecological cost, you can see the logic in budgeting for higher priced but likely more durable metal, wood, and glass options.

Build networks. By knowing your gardening neighbors, you'll have someone to share bulk purchases, equipment, and other resources. By knowing your local gardening-store owner and local seed producer, you have a basis for asking about plastic-free seed packaging and other plastic-less approaches.

Shop local. Local shopping reduces the packaging materials, such as plastic wrap and polystyrene peanuts, required to ship your items. This also improves your chances of sourcing plastic alternatives.

Buy in bulk. Plastic packaging becomes a larger issue as the package sizes get smaller. Think about a 1-gallon (3.8 L) jug of water versus 10 to 12-ounce (296 to 355 ml) bottles of water. While the volume of water is nearly the same, the gallon (3.8 L) jug contains less plastic than the sum of the individual bottles. The same applies to jugs of liquid soil amendments, which could be split among a group of gardeners. Having a truckload of woodchips or compost delivered does away with plastic packaging altogether.

Do your best to not *purchase* plastic. When your current plastic-based belongings wear out, be prepared to invest in nonplastic alternatives. For the plastic that's already in your life, use and care for it as best you can to prolong its life. If a free and useful plastic item comes your way, don't hesitate to use it. It's a far better option than sending it to a landfill.

Buy bark mulch, compost, leaf mold, and other soil amendments in bulk whenever possible.

GOODBYE, GAS POWER

AS A HUMAN LIVING in the twenty-first century, you were born into a fossil fuel–powered life. Even if you don't own a car and ride a bike everywhere, the bike probably came to your local bike store via a diesel-powered semi-truck and possibly by cargo ship using a diesel engine before that. While escaping fossil fuels isn't a reality for most Americans, reducing your reliance on them is possible.

Fossil fuels pose several issues. Fossil fuel extraction from land and ocean sources is harmful to host communities, meaning people, plants, and animals. Transport is dangerous, with an ever-present possibility of large and small spills and leaks. Pipelines built to move fossil fuels from their point of extraction to refinery and transportation hubs are equally as precarious, as well as disruptive to the natural and social communities they bisect. Processing facilities are governed by environmental guidelines but remain detrimental to the surrounding environment and human health. There's also the pollution that's released when the fossil fuels are burned off in engines and processed in industrial practices.

Modern reel mowers are easy to use and require no electricity or fossil fuels to cut the lawn.

BATTERY-POWERED EQUIPMENT

Fuel use in the garden is directly tied to the way you choose to manage your garden. Each year, Americans use about 1.2 billion gallons (4.5 billion L) of gasoline in their lawn mowers. (This figure doesn't include diesel fuel, used in some commercial mowing operations.) If building your ecological garden involves reducing the amount of lawn you have to mow in favor of native plantings, you're already on the right track to reducing your fossil fuel dependence. Even with some lawn left to mow or a fence line that a homeowners association requires you to keep clean, you can opt for battery- or human-powered equipment. (If you're looking for help in reducing the size of your turfgrass lawn, read chapter 2.)

Lawn mowers and weed trimmers powered by lithium-ion batteries have come a long way since their introduction. It used to be that their power wasn't equal to that of a gas model, rendering them impractical in less-than-ideal mowing conditions. Today's battery-powered garden equipment may even offer more power than some gas models.

The batteries charge using electricity—maybe yours comes from a renewable source—and their components include plastic parts and rare metals. Whether switching from gas power is the right choice for you comes back to the ecological gardening priorities you developed in chapter 1.

Batteries are generally not interchangeable between brands, but to have multiple pieces of equipment from the same brand means you can own fewer batteries (and chargers). For example, a chainsaw, weed trimmer, and lawn mower from the same manufacturer can all use the same battery.

HUMAN-POWERED EQUIPMENT

When getting your ecological garden started, you may opt for equipment with some oomph behind it to till the soil, dismantle structures, or move landscaping features. After this initial effort, your garden can be managed by human power.

Because you're relying on your own body to do this work, the right tool for the job will save your muscles, time, and energy. Select your tools with ergonomics and purpose in mind. Examples of human-powered garden equipment include:

- **Garden hoes.** Continually tilling the soil disrupts the soil web of life, so you don't need to have a tiller in your garden shed. Instead, a few types of hoes will do for spot weeding. A good arsenal may include a claw hoe to loosen soil, a diamond hoe to cut off weeds just below the soil surface, and a chopping hoe to go after large weeds with stubborn roots.

- **Shovels.** A shovel is not just a shovel. Different types of shovels have been designed for different uses. You may find yourself adding to your shovel collection as you develop your garden. Carving out a swale for water movement is simplest using a trench shovel or ditch shovel. Digging a hole to plant a tree requires a spade. Moving mulch can be done with a flat-bottomed shovel, but you might prefer using a steel-tine pitchfork or bedding fork.

- **Trimmers.** Instead of a gas- or battery-powered weed trimmer, a small area can be managed with hedge shears. Without major tree limbs to manage, a chainsaw can easily be replaced with a handsaw or tree pruners.

When possible, use human-powered tools to maintain your garden, such as manual hedge trimmers, edgers, and saws, rather than relying on gas-powered options.

- **Lawn mower.** A reel mower is an inexpensive, walk-behind, human-powered alternative to a gas or battery mower. This mower has a series of curved blades on a cylinder that turns as you push it, shearing the blades of grass in its path. On lawns kept short and maintained on a schedule, these mowers are easy to use. If you tend to have a less regimented mowing schedule, experience weeks at a time with rain that keep grass wet during the mowing season, or have an uneven yard, you could become frustrated with the reel mower's performance.

Greenwashing

The "living lightly" idea centers on less clutter and less waste, but as a person living in modern society, it's nearly inevitable that you'll have to purchase some new items. In making decisions that move you closer to being more ecologically minded in your garden and in other areas of your life, you'll run into confusing labels and messaging from manufacturers of "green" products who are trying to capitalize on your ecological interests. Greenwashing happens when companies use misleading claims about the environmental friendliness of their products and services. One example is for a product to be labeled as "made from recycled materials," while only 10 percent of the product's materials were actually recycled.

It's hard to protect yourself from greenwashing when labels and claims are not strictly regulated. The U.S. Federal Trade Commission has marketing guidelines in place, but they can't investigate every marketing claim made by every company. Your best defense against greenwashing is to do your own research into the company, their products, and the alternatives.

The BIGGER *Picture*

THERE'S A LOT OF INFORMATION in this chapter about taking your ecological gardening efforts one step—or many steps—further into a more holistic way of living as a responsible member of the natural world. Whether you're new to this way of thinking or using this as a refresher, it's easy to get overwhelmed; it's easy to shut down in light of how far the modern American lifestyle has drifted from our natural roots.

Every step you take toward a more ecological future has a ripple effect. Do your best, and release your judgment of yourself and others as everyone picks their way through this process. When you're not sure what you can or should do next, come back to a few key takeaways:

Slow down. Cheap-and-easy is the default setting for those of us with hectic lives. The rat race has convinced us that this is the only way. A more ecological approach to gardening—and to life—requires you to see through this myth. We have this to learn from traditional and Indigenous lifeways. Think about the people who stewarded the land in your neighborhood for thousands of years before streets were paved and pipelines were dug. We can't go back to that time, but we can learn from it.

Buy less. All around you are messages that you need more. See through the hype. Of course you want the newest/coolest/best thing to ever come across your screen. Do you need it? Will your life be better because of it? Will your ecosystem? Employ your critical thinking, research, and reasoning.

Buy responsibly. When you do decide you need to make a purchase, think about its life cycle. Its materials, the people and processes that made it, and what will happen to it after its useful life is over are all considerations that go into responsible purchasing decisions. Look at the labels.

Be creative and upcycle whenever possible. This clever gardener is using an old apple crate and metal colander as garden containers.

You may find the more responsibly made options are also the more costly options. Balancing your finances to match your values often means you end up buying less out of necessity.

Buy secondhand. While "vintage" comes in and out of style, buying used is always a good idea. Your neighborhood's Nextdoor group, Facebook Marketplace, Freecycle, and Craigslist are all online options for finding what you need. Most communities have thrift shops, which are often run by nonprofits and somehow benefit their missions. At least one national outdoors store has a secondhand program, both in their stores and online. Secondhand doesn't mean dirty or damaged. You'll be surprised what you can find, from clothing with tags still attached to attractive planting containers to well-made furniture.

Your secondhand purchase doesn't require the use of additional resources to manufacture, probably doesn't have to travel far to get to you, and doesn't arrive with unnecessary packaging. And it's costing you less than buying new.

Be careful with this tenet of conscious consumerism, though. Because you found it secondhand doesn't mean you *should* buy it. Go back to ideas at the top of this list: Slow down, and buy less.

Buy in bulk. You read about buying garden materials, like compost, in bulk. On a different scale, many of the ingredients you use in your home, and especially your kitchen, are available to purchase in bulk quantities. This doesn't mean you need to purchase fifty pounds (22 kg) of black beans. A "bulk" purchase could mean enough to fill a quart jar. The idea is that you're buying the items you need without the packaging that you don't need.

Reach for reusables. Think about the household items that you use and throw away: coffee cups, paper towels, deodorant tubes, and spray bottles of cleaning products, to name a few. Replace these with a reusable mug, cloth towels, a refillable deodorant, and refillable cleaning products. These are just some ideas to get your mind turning.

Use less-toxic cleaning alternatives. Keeping a clean environment is vital for good health, but we've been sold the idea that we need toxic chemicals to keep germs at bay. Look for cleaning products that are ecologically safer. Fragrances, irritants, flammable ingredients, and volatile organic compounds (VOCs) are harsh chemicals that can cause human and environmental harm.

Raised beds constructed from used pallets have created this beautiful backyard vegetable garden. Be mindful of the resources you use in your gardening practices and life.

Secondhand outdoor furniture is a fun addition to any garden.

Eat at home. An interest in cooking and growing more of your own foods may be a reason you're exploring ecological gardening to begin with. Cooking your own meals allows you to be in control of food waste, make healthier food choices, and do away with packaging from processed foods and takeout containers.

Working within a circular economy, turning your attention to more local resources, reducing your plastic and fossil fuel use in the garden, reducing your electrical and water consumption everywhere, and really thinking about your needs versus your wants are all part of an ecological outlook. As your vision for an ecological garden comes together, so will your plan for how these principles factor into the rest of your life. This is yet another example of how all things are interconnected. The changes you make in your backyard have the potential to go farther than you realize.

CONCLUSION

Create a garden that is a gift to others—human and otherwise.

Whether large or small, an ecologically sound garden has so much to offer to you, to other living beings, and to the Earth as a whole.

HERE IN THE ECOLOGICAL GARDEN, it's all about the community. This is a wide-ranging collection of living beings, including your neighbors (even those who have different ideas about how you should be using your yard), other gardeners, and garden-related local business owners and educators. Your community is also in the soil, with the microbial life and invertebrates working tirelessly to improve the soil web. It's among the existing plants and trees of your neighborhood, the birds and wildlife that are drawn to your garden, and insects, both friend and foe. Remember that list you made of all of the creatures you could find in your garden in chapter 3? Those creatures are important. The ecological community is inclusive of humans, not subservient to them.

You have the opportunity to create a garden that's a treasure for this community. Learn about its needs, and act on them in a way that's sustainable, regenerative, and every other ecology-related buzzword you can think of. Design a space that's right for you but doesn't exclude the other community members. Take the time to observe, both as a means of enjoying your space and of learning about it. Wait for the butterflies to show up, and pat yourself on the back for remediating your landscape when they do.

Reflect on the world around you and how your garden fits into it. Reflect also on why you've picked up this book in the first place: What is it that attracts you to the life of an ecological gardener? That reason makes a difference in how you approach this practice.

There has never been a more pressing need for a gardening book to guide us through growing plants in uncertain climates. This book is a step toward preparing you for a more ecologically minded future.

Whether you're a long-time conventional gardener or totally new to working with plants and landscapes, there's a place in ecological gardening for you. Conventional gardening strategies and garden design aren't totally in line with the more holistic outlook, but neither are they completely at odds.

Opting to create a diverse, biologically rich garden is something you can feel good about.

Don't be fooled into thinking an ecological garden is a wildly unmanaged space. You're still managing the garden, this time with a holistic approach. Your ecological garden can maintain the size, shape, texture, and color design guidelines that have been long accepted by garden professionals. If you're willing to make the right compromises, you can combine your personal ecological gardening principles with more traditional garden rules, and your neighbors might not even know the difference.

It's your principles, which you outlined in the first chapter of this book, that will guide you in the design process, from your plant selections to your mulch choices and the ways you manage water in and around your garden space. Take the time now to go back to page 20 and look at your ecological-gardening priorities. These principles and the notes you made about them will help guide your decision-making throughout the gardening season. As you grow as a gardener, it's important to revisit this list so your practices can evolve with your knowledge and experience.

When it comes time to select your plants, learn about your options and reach for those that are most right for your climate and, more specifically, your yard. Too often, we plant things that gardeners always plant because we receive messaging about the way things should be done. Browse the seed catalogs and nursery aisles, virtually or in person, and see what's available. Compare those options to what your climate and community actually need.

We get into trouble as gardeners when we design spaces and choose plants that are counter to what nature wants. Gardening becomes work instead of joy when we spend all of our time watering, replacing plants, and tilling ground. An ecological approach puts the right plant in the right place and ultimately reduces our time spent toiling.

Native plants and those most suited to the growing conditions outside your door are best for your natural neighbors and water-conservation concerns. Let your plants work for you in preventing erosion, shading your home, and more. Choose a diversity of plants that will provide flowers for pollinators throughout the seasons, habitat for wildlife in winter, and maybe a little something for you to eat too.

Respect water, our precious resource, by creating a water-wise garden. While many parts of this country are experiencing unheard-of rainfall events or crippling drought, and sometimes one followed by the other, gardeners can call upon their ecological know-how to create spaces that help to mitigate the effects. Your landscape changes the way water traverses the land. The topography of the area surrounding you is not in your control, but that of your own yard is. Berms, swales, hügelkultur mounds, green roofs, and rain gardens are all ecologically minded solutions to rainfall problems.

While homeowners across the country are watering their nonnative turfgrass lawns to maintain a certain aesthetic, groundwater reserves and drinking-water sources are being stretched to the limit. The opportunity presents itself here to reduce the size of your monoculture lawn (and the chemical use, mowing, and watering it requires) or do away with it altogether in favor of a more biologically diverse, less resource-hungry garden.

It's a fact that all living beings require water to survive, but some more than others. Responsible garden irrigation comes back to plant choice as well as garden design. The ecological gardening design may not align with more conventional garden design in this way.

As your garden comes together, from concept to installation and into its maturity, you have maintenance tasks ahead. Because a primary aim of ecological gardening is to build the soil, your need for fertilizers is greatly reduced. It's possible you'll really only need them in the case of heavy-feeder vegetable plants and those in containers. For the most part, a healthy soil will provide the nutrients your plants need. You'll achieve this using soil amendments, including compost. Because of compost's value to your garden and to the environment overall, you yourself will be making compost, maybe in a bin in your backyard, in a vermicompost container under your kitchen sink, or through a municipal composting program.

Rely on the balance of nature to work out problems. This is far easier to do in a biodiverse ecological garden where the natural system of checks and balances is left to develop over time.

There will be weeds to pull, but the measure of what makes a weed a *weed* is different in an ecological garden. Before removing plants, you'll question their role in the ecosystem and in that particular spot in your garden. Mulch will suppress a lot of weeds while also maintaining soil moisture. There's no need for tilling here, so new weed seeds aren't being brought to the surface, eventually reducing the weed seed bank awaiting germination.

Insects that we call "pests" will eventually arrive, including the Japanese beetles snacking on your blackberries and aphids sucking sap from your milkweeds. A pesticide spray is the easy option, but it's not the ecological one, and how you choose to handle the pest population will depend on your principles. Ecological gardening asks you to wait it out, letting the predators take care of them (birds for beetles, ladybugs for aphids), and trusting that your soil is healthy enough to make your plants strong enough to withstand the attack.

Seasonal upkeep is less intense than it would be in a conventionally managed garden and yard, with dead and dormant plant materials left standing and fallen leaves put to work for their habitat and nutrient benefits. Dividing and propagating your own plants, plus saving seeds, gives you the pleasure of having something to share with others and keeps your garden plant population growing.

Having spent all this time learning about and considering the nuances of ecological gardening, it's likely you've noticed where these principles can apply to other parts of your life. As beings in the ecosystem, our actions have real impacts system-wide. A closed-loop system, in the garden and in our lives at large, runs counter to the linear system of production and waste that we've come to know in today's United States. Getting closer to a closed loop by sourcing our needs closer to home, reusing what we have, relying on fewer new purchases, and reducing waste are truly ecological acts. At this point, rampant plastic use comes into question. Plastic use in the garden is hard to escape, but the ideas presented in this book combined with your own resourcefulness is a start.

Knowing your place in the garden as a member of the natural community rather than as the one in charge of it is the starting point for a beautiful, thriving ecological garden. This humble approach pays dividends in the ecosystem services a garden like yours can provide.

Taking an ecological approach to gardening benefits your and your community's quality of life in a multitude of ways:

- Mitigating runoff and extreme rainfall events
- Easing the urban heat island effect and moderating temperatures
- Filtering pollutants from air, soil, and water
- Adding to the biodiversity of your region, both in terms of the plants you cultivate and the natural life they invite to the garden
- Offering a respite from the hectic built world around you (and bringing up property values to boot)
- Reducing noise, light, and visual pollution
- Giving you an outlet to express your ecological-living commitment

As you build and tend your ecological garden, remember nature's cycles and the balance present in a healthy ecosystem. It is a journey, and one well worth taking. Remember that your decisions in the garden are impacting many, and you have the power to be an incredibly positive force. You'll make mistakes along the way, but do your best to consider them learning opportunities and move on. Know that you are doing good work, to whatever level you are able, and that the decisions you make will now be beneficial to all of the natural world.

There's no doubt your neighbors will take notice of your efforts and some may even adopt a more ecological focus in their own landscape care. Healthy, diverse gardens have a way of drawing others in and inspiring them to follow a similar path.

JOIN THE AHS

TO JOIN THE AMERICAN HORTICULTURAL SOCIETY (AHS) and enjoy all the benefits membership has to offer, please visit our website ahsgardening.org/join or scan the QR code below for our membership information page.

All members receive an exciting lineup of benefits that share the joy of gardening and strengthen your skills and knowledge.

- Receive our award-winning magazine, *American Gardener*.
- Get free admission and privileges at more than 360 public gardens and arboreta throughout North America through the Reciprocal Garden Network.
- Enjoy discounts on educational programs, garden shows, seeds, books, and more.
- Discover opportunities to explore the world's finest gardens on trips with fellow plant lovers.
- Access members-only gardening resources on our website.
- And much more.

ACKNOWLEDGMENTS

THE AHS EXTENDS GRATITUDE to its Horticultural Advisory Council members who contributed their expertise during this book's review process:

Dr. James Folsom
Dr. Mary Hockenberry Meyer
William McNamara
Claire Sawyers

The AHS staff and supporting consultants involved in the review process for this book were: Suzanne Laporte, David Ellis, Courtney Allen, Mary Yee, Susan Friedman, and Dan Adler.

The AHS expresses thanks to the book's acquiring editor, Jessica Walliser, and to the team at Cool Springs Press.

PHOTOGRAPHY CREDITS

Alamy: pages 15, 35, 46, 95, 117, 140, 141, 154, 170, 185, 204, 211, 226

American Horticultural Society: pages 6, 7

Austin Hyler Day: pages 20, 21 (top), 23, 47, 50, 71 (top left), 92, 118, 120, 137, 170, 180, 199, 229

Christina Salwitz: pages 11, 34, 88 (top), 142 (bottom), 153, 158, 163, 194, 222

Janet Davis: pages 9, 24, 32, 39, 58, 67, 70, 84, 85, 86, 87, 89 (bottom), 102, 116, 130, 133 (top), 134, 136, 139, 142 (top), 168, 172, 186, 227, 240

Jennifer McGuinness: pages 64, 65 (left), 167, 206, 214

JLY Gardens: pages 16, 21 (bottom), 31, 37, 44 (top), 49, 55, 60, 61, 63, 65 (right), 68, 74 (bottom), 75, 76, 77, 78, 79, 80, 83, 89 (top), 93, 97 (top), 99, 100, 104, 105, 107, 108, 109, 110, 114, 121, 128, 131, 149, 159, 171, 175 (top), 176, 177, 178, 189, 191, 192, 193, 198, 200, 201, 203, 209, 212, 213, 219, 224, 230, 233

Kathy Jentz: pages 30, 36, 48, 148, 156, 157, 205

Kelly Norris: pages 14, 22, 26, 57, 70, 71 (top right, bottom), 101, 111, 126, 127, 133 (bottom), 138 (top), 150, 151, 179, 225

Noelle Johnson: pages 36, 129, 160, 161

Shutterstock: pages 4, 8, 17, 33, 37, 41, 44 (bottom), 71, 72, 73, 74 (top), 82, 88 (bottom), 93, 94, 96, 97 (bottom), 98, 112, 115, 119, 122, 138 (bottom), 145, 146, 152, 162, 164, 165, 173, 174, 175 (bottom), 181, 182, 183, 187, 190,196, 197, 202, 208, 210, 215, 216, 217, 218, 223

Stephanie Rose: pages 19, 38, 40, 51, 106, 220, 221

Susan Mulvihill: pages 12, 91

Tracy Walsh: pages 25, 29, 38, 113

INDEX

E

F

G

H

I

K

L

M

N

O

P

Q

R

S

T

U

V

W

X

Z